THE DYNAMICS OF

Heritage, Culture and Identity

Series Editor: Brian Graham,
School of Environmental Sciences, University of Ulster, UK

Other titles in this series

Landscape, Race and Memory
Material Ecologies of Citizenship
Divya Praful Tolia-Kelly
ISBN 978 0 7546 4957 1

Unquiet Pasts
Risk Society, Lived Cultural Heritage, Re-designing Reflexivity
Edited by Stephanie Koerner and Ian Russell
ISBN 978 0 7546 7548 8

Culture, Heritage and Representation
Perspectives on Visuality and the Past
Edited by Emma Waterton and Steve Watson
ISBN 978 0 7546 7598 3

Sport, Leisure and Culture in the Postmodern City
Edited by Peter Bramham and Stephen Wagg
ISBN 978 0 7546 7274 6

Valuing Historic Environments
Edited by Lisanne Gibson and John Pendlebury
ISBN 978 0 7546 7424 5

Southeast Asian Culture and Heritage in a Globalising World
Diverging Identities in a Dynamic Region
Edited by Rahil Ismail, Brian Shaw and Ooi Giok Ling
ISBN 978 0 7546 7261 6

Jewish Topographies
Visions of Space, Traditions of Place
Edited by Julia Brauch, Anna Lipphardt and Alexandra Nocke
ISBN 978 0 7546 7118 3

The Dynamics of Heritage
History, Memory and the Highland Clearances

LAURENCE GOURIÉVIDIS
Université Blaise Pascal, France

Routledge
Taylor & Francis Group
LONDON AND NEW YORK

First published 2010 by Ashgate Publishing

Published 2016 by Routledge
2 Park Square, Milton Park, Abingdon, Oxfordshire OX14 4RN
711 Third Avenue, New York, NY 10017, USA

First issued in paperback 2016

Routledge is an imprint of the Taylor & Francis Group, an informa business

British Library Cataloguing in Publication Data
Gouriévidis, Laurence.
The dynamics of heritage : history, memory and the Highland Clearances. -- (Heritage, culture and identity)
1. Forced migration--Scotland--Highlands--History--18th century--Exhibitions. 2. Forced migration--Scotland--Highlands--History--19th century--Exhibitions.
3. Highlands (Scotland)--Emigration and immigration--History--18th century--Exhibitions. 4. Highlands (Scotland)--Emigration and immigration--History--19th century--Exhibitions. 5. Historical museums--Scotland--Exhibitions--Evaluation.
I. Title II. Series
333.3'14115'07411-dc22

Library of Congress Cataloging-in-Publication Data
Gouriévidis, Laurence.
The dynamics of heritage : history, memory and the Highland Clearances / by Laurence Gouriévidis.
p. cm.
Includes index.
ISBN 978-1-4094-0244-2 (hbk.)
1. Land tenure--Scotland--Highlands--History--19th century. 2. Eviction--Scotland--Highlands--History--19th century. 3. Museums--Scotland. 4. Collective memory--Scotland. 5. Cultural property--Scotland. I. Title.

HD615.G68 2010

2010017414

ISBN 13: 978-1-138-24609-6 (pbk)
ISBN 13: 978-1-4094-0244-2 (hbk)

Contents

To my parents and to the memory of my grandparents.

List of Figures

List of Tables

The ill-fated Clearances [are] one of the saddest chapters in Highland history, and one which, I am afraid, the young Gael of today has completely forgotten, because I am confident his attention has never been drawn to that lurid epoch. How many children in Sutherlandshire schools today could write a one-page essay on the Strathnaver Clearances at the beginning of last century?

The Clarion of Skye, November 1952 (22): 5

Like history, memory is inherently revisionist and never more chameleon than when it appears to stay the same.

(Raphael Samuel 1994: x)

Preface

To state that the Highland Clearances is one of the most contentious issues in Scottish history and culture, stirring impassioned reactions and emotions is hardly novel and unusual (Lynch 1992: 367-70, Richards 2000: 1-14). Not only were they controversial in their times, as indicated by the conflicting body of evidence surrounding them, but the debates that they continue to provoke amongst historians and more widely are also testimony to the hefty emotional and cultural baggage with which they are still encumbered. It makes their representation in such public spaces as museums and heritage centres all the more problematic for the individuals involved, and all the more interesting for any study related to the memorialisation of the past and its implications. The purpose of this book is both to explore the place of the Clearances in Scottish heritage – in its most capacious socio-cultural sense – and the reading and interpretation of the period presented in Scottish museum-style venues.

That Highland symbols have enjoyed more than an inordinate part in the construction of Scotland's imagery and heritage needs little demonstrating – its sartorial iconography, or its haunting and majestic landscape are just two examples of a phenomenon which gathered momentum at about the same time as the Clearances were in full swing (Prebble 1988, McCrone et al. 1995: 4-8, Clyde 1995). The romantic appeal of the region's purported 'wilderness' grew as fast as it was haemorrhaging its people (Withers 1992, P. Macdonald 2009); concurrently attitudes towards land use in the region became strongly determined by either a utilitarian or a scenic and, more recently, scientific vision of the value the land might have (Smout 1993). Mythical images thus mingled with more ideologically-based critiques of policies of social engineering, landowning power and, in modern parlance, ethnic cleansing. As a result, readings of nineteenth-century highland and island history and culture are often bunched up around sets of issues highlighting binary oppositions such as Improvement and Romance, Clearance and Improvement or landlords and crofters, reflecting competing contemporary discourses and visions (Hunter 1976, Womack 1989).

The intellectual and economic bases of the Clearances are clear enough. Eighteenth- and nineteenth-century Scotland was largely driven by the concept of Improvement which was a byword for economic development – agricultural rationalisation and transformation leading to increased productivity, and industrial innovation and advance. For the Highlands and Islands, this implied the complete restructuring of their economy with comprehensive agrarian reforms, such as changes in land use and introduction of large-scale sheep-farming, coupled or not with the establishment of new industrial ventures, more specifically kelping, fishing and textiles. The combination and timing depended on estate management and geographical or environmental specificities. Such changes also implied severe social disruption as peasant communities were displaced or removed from the land they occupied. So between the middle of the eighteenth century to 1886, when the Crofters' Act bringing evictions to an end was passed, after a period of social and political turbulence known as the Land Wars, thousands of families were 'removed' or 'cleared' off the land. Some were relocated elsewhere on their landowners' estates, some evicted outright, migrating to the expanding industrial towns of the central belt, or emigrating to other parts of the British Empire.[1]

As a portmanteau term, 'Clearances' does little justice not only to the type of displacement involved – forced removal accompanied either with emigration or resettlement, or direct expulsion (Richards 1982: 5-6) – but also to the diversity and multi-facetted nature of a process, which was a succession of episodes differing in terms of scale, chronology, geographical location, social and economic context – such as downturn in economic activity or famine crisis – proprietorial situation and orientation, and of course effects, to name but a few of the issues that the term conceals, and yet, require clarification if anything but sweeping and oversimplified statements is to be produced. On this historians agree. Individual Clearances are associated with phases and are to be linked with specific estate policies, the nature and orientation of which are, in differing degrees, a matter of environmental, socio-economic and political context or personal propensity and means of individual proprietors (Richards 1982, Macinnes 1996, Devine 1994a). This atomised nature of the Clearances, however, is also what makes their representation challenging for museums which find themselves in the horns of a dilemma, arguably less so if they concentrate on a very particular locale and period. They have to capture and retain the attention of visitors with a range of assumptions and varying levels of knowledge concerning the subject, in exhibitions where space is often at a premium and where information has to be pithy without resorting to simplifying generalisation.

The narratives of the Clearances constructed in museums along with the strategies of communication used are one aspect of this book's analysis, however, museums are located in time and place and those narratives and their formats have also been affected by an array of factors: changes in museum practice and thinking, economic pressures such as the impact of tourism and heritage on regional development, and the cannibalisation of the past for political or ideological purposes, notably in processes of identity construction. Such aspects also form a substantial part of the book's argument. In Laurajane Smith's words:

> Heritage can be understood as an important political and cultural tool in defining and legitimizing the identity, experiences and social/cultural standing of a range of subnational groups as well as those of the authorizing discourse. However, it

1 See Chapter 2 for historiographical details.

> may also be an important resource in challenging received identity and cultural/social values (L. Smith 2006).

So museum practitioners are ultimately in the business of making meaning out of the past often in the light of present conditions and concerns.

The book's argument relies on the study of 15 Scottish museums; two are national institutions; one is run by a local council and the remainder are independent. Each of these museums was visited, in some cases several times over the course of the past twenty years, enabling a diachronic perspective and some sense of the evolution of the treatment of the period and of museological practice in the late twentieth century. 'Museums' is taken as an umbrella term, as given the changes in methods and communication strategies which have affected the profession since the sixties, they have come to rely on a variety of technological means, rather than simply the collection and exhibition of artefacts, in order to impart a sense of the past to visitors. The venues surveyed are listed in two tables. Table P.1 outlines the general characteristics of each museum, whilst Table P.2 lists more specific aspects linked with the interpretation of the Clearances. Venues are presented in chronological order of opening, though some went through refurbishment affecting the nature of their presentation of the period, thus making it possible in some cases to chart changes and trends, and establish comparisons. Some of the exhibitions included here no longer exist or were in the process of being altered at the time of writing. For instance, *Aros*' exhibition lasted from 1993 to 2002, *Co Leis Thu?*'s from 1991 to 2000 and after complete makeover and change of location, the venue was renamed *Seallam!* Visitor Centre. The exhibition referred to at Timespan ended in 2007. The Museum of the Isles started integrating a more exhaustive treatment of the Clearances in 2002 and my observations will briefly contrast the 1984 and 2002 exhibitions.

In addition to visits, a questionnaire was devised for the independent museums surveyed, which ten of them agreed to answer (see Appendix). The questions covered a range of issues and themes, the first part of these dealing with the background, nature and objectives of the venues, the second with the Clearances and the implications of their interpretation. The answers received, some of them fairly extensive, provided invaluable data used throughout the argument.

The Dynamics of Heritage is concerned with the memorialisation of the Highland Clearances over the twentieth century, examining the interaction between history and memory and focusing on museums as a memory medium. This book, whilst making no claims to provide a comprehensive analysis of the memory and legacy of the Clearances, endeavours to bring together a range of trans-disciplinary themes and concepts such as collective memory, heritage and more precisely museum development, and the function of the past and heritage in definitions of identity. It approaches these issues from the perspective of social history looking at the structures, groups and individuals involved in the representation of the period for public consumption in Scotland. Given the vast expansion of the field of memory in the social sciences over the past decades, the conceptualisation of its approach requires more than a few words and the first chapter is dedicated to providing a framework for the book's methodology and argument, and ends with a presentation of its structure and content.

Laurence Gouriévidis
March 2010

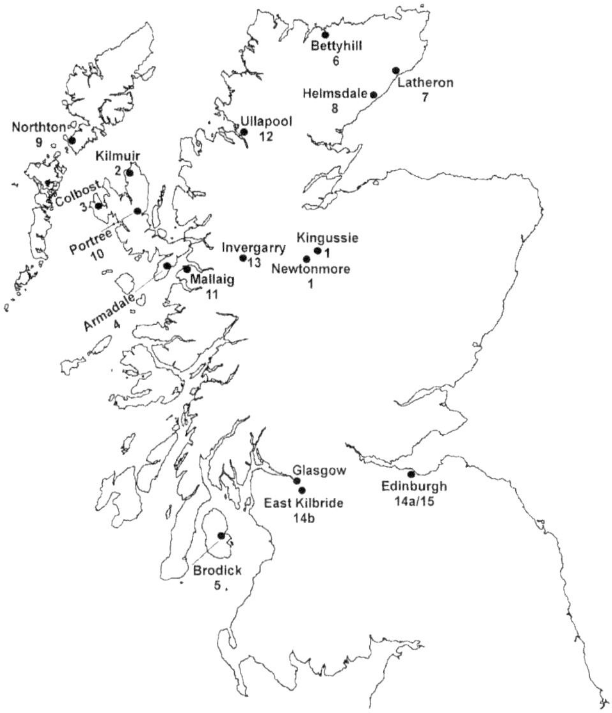

	Name	Location
1.	Highland Folk Museum	Kingussie, Strathspey (1944)
		Newtonmore, Strathspey (1996)
2.	Skye Museum of Island Life	Kilmuir, Isle of Skye (1965)
3.	Colbost Folk Museum	Colbost, Isle of Skye (1969)
4.	Museum of the Isles (Clan Donald Centre)	Armadale, Isle of Skye (1972)
5.	Isle of Arran Heritage Museum	Brodick, Isle of Arran (1976)
6.	Strathnaver Museum	Bettyhill, Caithness (1976)
7.	Clan Gunn Heritage Centre	Latheron, Caithness (1985)
8.	Timespan	Helmsdale, Sutherland (1987)
9.	*Co Leis Thu?*	Northton, Isle of Harris (1991-2000)
	Seallam! Visitor Centre	Northton, Isle of Harris (2000)
10.	*Aros*	Portree, Isle of Skye (1993)
11.	Mallaig Heritage Centre	Mallaig, Lochaber (1994)
12.	Ullapool Museum and Visitor Centre	Ullapoool, Wester Ross (1988)
13.	Glengarry Heritage Centre	Invergarry, Lochaber (1999)
14a.	Scottish Agricultural Museum	Ingliston, by Edinburgh (1980-2001)
14b.	National Museum of Rural Life	Wester Kittochside, East Kilbride (2001)
15.	Museum of Scotland	Edinburgh (1998)

Figure P.1 Location of museums/heritage centres

Table P.1 General characteristics of museums (in chronological order of opening, national in separate section)

Name	Opening date	Location	Type	Focus	Survey	Visit(s)	Facilities
1. Highland Folk Museum	1944	Kingussie Strathspey	Privately owned: I.F. Grant From 1974 run by local council	Folk life – regional		1992	
	1996	Newtonmore Strathspey	Highland Council	Living history: Regional 1700s township 1930s buildings		1996 2000 2007	Café Gift shop
2. Skye Museum of Island Life	1965	Kilmuir Skye	Independent owner: Jonathan MacDonald	Folk life Skye history in ceilidh house mainly from modern times	✓	1992 1996 2003 2007	Gift shop
3. Colbost Folk Museum	1969	Colbost Skye	Independent owner: Peter MacAskill	Folk life		1992 2007	
4. Museum of the Isles	1972 1984 2002	Armadale Skye	Independent Clan Donald Lands Trust	Clan Donald history	✓	1992 1996 2003 2007	Restaurant Gift shop Archive/ Library
5. Isle of Arran Heritage Museum	1976	Brodick Arran	Independent Arran Heritage Trust	Folk life Arran general and natural history mainly from modern times	✓	2008	Tea room Gift shop Archive
6. Strathnaver Museum	1976	Bettyhill Caithness	Independent Strathnaver Museum Trust	Strathnaver history with emphasis on Clearances Clan Mackay	✓	1992 1994 2007 2008	Gift shop

Table P.1 continued

Name	Opening date	Location	Type	Focus	Survey	Visit(s)	Facilities
7. Clan Gunn Heritage Centre	1985	Latheron Caithness	Independent Trust	Clan Gunn history	✓	1992 2007	Gift shop Archive
8. Timespan	1987-2006	Helmsdale Sutherland	Independent Helmsdale Heritage Society	General and natural history of Helmsdale and surrounding area in particular the Strath of Kildonan	✓	1992 1994 2008	Gift shop Café Art Gallery Archive
9. *Co Leis Thu?*	1991-2000	Northton Harris	Independent: Initiated by Bill Lawson and made over to Northton Heritage Trust in the late 1990s	History of Harris from pre-history		1997	Gift shop Genealogical centre
Seallam!	2000			Hebridean history from pre-history	✓	2008	Art gallery Learning centre for UHI
10. *Aros*	1993	Portree Skye	Independent Private owners	Skye's story mainly from modern times		1996 2003 2007	Restaurant Gift shop Art gallery, auditorium
11. Mallaig Heritage Centre	1994	Mallaig Lochaber	Independent Mallaig Heritage Trust	Heritage of Mallaig and West Lochaber from modern times	✓	2007	Gift shop

Table P.1 continued

Name	Opening date	Location	Type	Focus	Survey	Visit(s)	Facilities
12. Ullapool Museum and Visitor Centre	1988 1996*	Ullapool Wester Ross	Independent Ullapool Museum Trust	General and natural history of parish of Lochbroom	✓	1992 1997 2008	Gift shop
13. Glengarry Heritage Centre	1999 2006*	Invergarry Lochaber	Independent Trust	History of Glengarry district from modern times Genealogical emphasis	✓	2007	Café Resource centre
14a. Scottish Agricultural Museum	1980-2001	Ingliston Edinburgh	National	Scottish country life		1992	
14b. National Museum of Rural Life	2001	Wester Kittochside East Kilbride		Scottish country life from c. 1750 Working farm		2008	Café Gift shop Study store Learning centre
15. Museum of Scotland	1998	Edinburgh	National	Scotland's culture and history		1998 2003 2007 2008	Restaurant Gift shop Etc.

Note: For opening date, when several dates are mentioned, they indicate that the venue underwent refurbishment or transformation. The survey can be found in Appendix; * In present form.

Table P.2 Clearance narratives in museums (in chronological order of opening)

Name	Inclusion	Media	Sources used	Viewpoint	Survey	Visit(s)
1. Highland Folk Museum – Newtonmore	1996-?	• Text • Featured in introductory video				1996 2000 2007
2. Skye Museum of Island Life	From the 1980s in 'ceilidh house'	• Text • Photographs/sketches	Not specified	Crofters	✓	1992 1996 2003 2007
3. Colbost Folk Museum	✓	• Text – newspaper clippings and brief introductory note				1992 2007
4. Museum of the Isles	1972: ? 1984: Brief mention in chronological table 2002	Since 2002: • Text • Images • Map (emigration routes) • Model: ship • Artefacts	• Modern • Contemporary	Partisanship avoided	✓	1992 1996 2003 2007
5. Isle of Arran Heritage Museum	Temporary in the late 1970s and in 2004-2005 Permanent from 2006	• Artefacts • Text • Images • Specific video • Local maps and tables (Arran and Canada)	• Contemporary	Islanders/emigrants	✓	2008
6. Strathnaver Museum	✓	• Artefacts – archaeological • Text • Images • Local maps and tables • Trail to cleared townships	• Contemporary • Oral testimonies	Have tried not to take sides	✓	1992 1994 2007 2008

Table P.2 continued

Name	Inclusion	Media	Sources used	Viewpoint	Survey	Visit(s)
7. Clan Gunn Heritage Centre	✓	• Text • Images • Models: shepherd + emigrant woman • Featured in general introductory video	• Contemporary	Crofters	✓	1992 2007
8. Timespan	✓	• Text • Images • Models and sound-effects: eviction scene + landlords • Featured in exhibition video	• Contemporary • Oral history • Modern	Crofters	✓	1992 1994 2008
9. *Co Leis Thu?* *Seallam!*	✓	• Text • Photographs	• Contemporary • Modern	Crofters	✓	1997 2008
10. *Aros*	1993-2002	• Text • Images • Models: emigrant ship/Johnson & Boswell/Napier commission • Part of commentaries and effects in multi-lingual headset system • Featured in exhibition video				1996 2003 2007
11. Mallaig Heritage Centre	✓	• Text • Images • Local maps, tables & graphs • Brief mention in exhibition video	• Contemporary • Modern	Crofters and landlords	✓	2007

Table P.2 continued

Name	Inclusion	Media	Sources used	Viewpoint	Survey	Visit(s)
12. Ullapool Museum and Visitor Centre	✓	• Text • Images • Featured in introductory video	• Contemporary • Modern • Oral history	Crofters	✓	1992 1997 2008
13. Glengarry Heritage Centre	✓	• Text • Images • Maps and tables • Trail to cleared township	• Modern • Contemporary	Emigrants	✓	2007
14b. National Museum of Rural Life	✓	• Text • Images • Specific video • Artefact				2008
15. Museum of Scotland	✓	• Text • Images • Featured in Scotland & the World video • Artefacts: miscellaneous Clearance memorabilia				1998 2003 2007 2008

Note: The information in 'sources used' and 'viewpoint' is based on survey answers (see Appendix for survey).

Acknowledgements

The broad theme of this book emerged from my research in the late 1980s, which dealt with the image of the Clearances in history, literature, politics and museums. Those early seeds took a long time to germinate and eventually came into bloom during a most inspiring visiting fellowship which the Australian National University, Canberra, granted me in 2005. Discussions with colleagues specialising in museology and memory studies along with museum curators, prompted me to explore further the museum section of my work.

Over time I have amassed many debts of gratitude. First and foremost, my warmest thanks go to Chris Smout who has ploughed through this book's draft and whose comments and observations have been both useful and thought-provoking. However, all failings in the end product are my own. Thanks also to Margaret Richards for producing the map that locates the venues surveyed in this book.

As mentioned my stay at the Australian National University was invaluable. Exchanges with scholars based at the then Humanities Research Centre, with other visiting fellows and with conference/seminar participants have helped me fine-tune my thinking in terms of conceptualisation and methodology. In particular I wish to thank Silke Arnold-de Simine, Laurence Brown, Mary Hutchison, Kylie Message, Paul Pickering, Susannah Radstone and Alex Tyrrel.

Field work in museums was carried out over many years as I visited and re-visited a number of local and national museum spaces in Scotland and Australia. I am greatly indebted to staff and volunteers working in the venues surveyed in this book, which I hope is also clear evidence of their dedication and enthusiasm. Without their help and cooperation, my research would have been impoverished. Particular thanks are extended to those who took time to answer the questionnaire and enabled me to use archival material related to their venue. I am specially grateful to Viv Szekeres, Director of the Migration Museum in Adelaide between 1987 and 2008 for sharing her thoughts and writing on the development of the museum she helped run, and to the members of the Scottish Associations of South Australia met in 2005. Thanks also to Hugh Cheape for helpful conversations on Highland material culture and the making of the Museum of Scotland.

Much of my more recent fieldwork was supported by the Strathmartine Trust with the award of a grant in 2007 that enabled me to travel to the Highlands and the Hebrides in order to update and further my research.

This project would have been greatly postponed had I not had time and freedom from teaching and administration and I wish to thank my colleagues for their support during the research leave that Blaise Pascal University granted me in 2008-2009. I am all the more grateful to them and BPU as leave of absence is something of a rarity in the French university system.

Finally and most importantly, I want to thank my parents and my partner for their unwavering support, encouragement and good humour, my father and Reg also provided technical help with photographs and presentation.

List of Abbreviations

ASVA	Association of Scottish Visitor Attractions
HIDB	Highlands and Islands Development Board
HIE	Highlands and Islands Enterprise
HIES	Highland and Island Emigration Society
ICOMOS	International Council on Monuments and Sites
MSC	Manpower Service Commission
MJ	*Museums Journal*
NT	*Northern Times*
PJ	*Press and Journal*
SAoSA	Scottish Associations of South Australia
SCLMT	Scottish Country Life Museum Trust
SMC	Scottish Museums Council
SMN	*Scottish Museums News*
SNH	Scottish Natural Heritage
UNESCO	United Nations Educational, Scientific and Cultural Organisation
WHFP	*West Highland Free Press*

Museums surveyed or mentioned

Arran HM	Isle of Arran Heritage Museum
Colbost FM	Colbost Folk Museum
Glengarry HC	Glengarry Heritage Centre
Gunn HC	Clan Gunn Heritage Centre
HFM	Highland Folk Museum
IM	Immigration Museum (Melbourne, Australia)
Mallaig HC	Mallaig Heritage Centre
MM	Migration Museum (Adelaide, Australia)
MoI	Museum of the Isles
MoS	Museum of Scotland
NMoAoS	National Museum of Antiquaries of Scotland
NMoRL	National Museum of Rural Life
Seallam! VC	*Seallam!* Visitor Centre
SMoIL	Skye Museum of Island Life
Ullapool M&VC	Ullapool Museum and Visistor Centre

Chapter 1
Memory, History and Museums

> *Memory, history, and relics offer routes to the past best traversed in combination. Each route requires the others for the journey to be significant and credible. Relics trigger recollection, which history affirms and extends backward in time. History in isolation is barren and lifeless; relics mean only what history and memory convey ... Significant apprehension of the past demands engagement with previous experience, one's own and others', along all three routes* (David Lowenthal 1985: 249).

Collective memory, politics and identity

Past decades have witnessed a profusion of works in the social sciences tackling the subject of 'memory' and the notion has been conceptualised and unpacked in ways which extend beyond the individual and encompass the collective. 'Collective memory' has been construed in subtly varying ways and is clad in many adjectival garments often marking disciplinary boundaries: public memory, social memory, cultural memory and to make it operative in the context of this study some defining has to take place. The starting point of reference is Maurice Halbwachs' seminal demonstration that the memory of individuals is shaped by their belonging to definite social groups which provide them with the means of understanding the past; individual memories are to a large extent socially determined weaving together personal remembrances and a perception of the past which is the result of influences that are social in nature. Using as illustration kinship, religious and class-based groups, Halbwachs showed that individual memory relies on the 'social frameworks' of those constituent groups and that memories tend to be forgotten when the groups that used to sustain them disappear; collective memory and individual memory cannot be easily separated, individual memory being affected by changes in relationships to collective milieus or by the transformation of these milieus. Halbwachs also foregrounded the importance of space both as a shaping force in social relations and as a medium for the transmission of memories (Halbwachs 1980, 1992, 1994, 1997). His most historical work, *La topographie légendaire des évangiles en terre sainte: Étude de mémoire collective* (2008), is the empirical implementation of these theoretical views; it explores the making and altering of the landscape of the Holy Land through a survey of its sacred sites and brings to light the political dimension of the process with the increasingly powerful role of Christianity. Although Halbwachs' theory has been altered and considerably refined by subsequent scholars, his thinking is the cornerstone of a vast number of studies. Not surprisingly historians concerned with oral testimonies often refer to his work in their analyses of the wider social frameworks structuring individual recollection in terms of content and form (Thompson 1988, Tonkin 1992, Fentress and Wickham 1992: 87-143). It is also influential in research looking into the public institutions which construct the past to be remembered, for instance Holocaust monuments or commemorative displays in such official sites as the United States Capitol (Schwartz 1982, Young 1993). Others still have sought to complement his work by studying the role of ritual performances in processes of collective remembering focusing on commemorative ceremonies and bodily practices (Connerton 1989).

What is widely admitted now is that just as individual memory is constructed so is the memory of groups. Very broadly, social psychologists who have examined the subject of collective memory have dissected the link between individual mental processes and remembering as a social act whilst historians have explored the political and economic forces at work in memory construction. It follows that the ways in which historians have conceptualised their study of collective memory have not concerned its accuracy but its social dimension and the manner in which it is exercised. They have identified the 'memory carriers', or the sites where it is crystallised, the processes which secure its transmission and its cultural, economic and political ramifications. Their approach underscores memory as the result of social interaction and as a locus for conflict, negotiation, reinterpretation conditioned by present needs and priorities; the struggle for the control of memory finds its roots in current disputes of a socio-economic, cultural and political cast (Thelen 1989).

French scholarship offers compelling examples of analyses of such processes. Halbwachs' demonstration acts as the foundation stone of a vast edifice recently expanded by the monumental collective work led by Pierre Nora, *Les Lieux de mémoire* (1997a), and the philosophical reflections of Paul Ricoeur (1998a, 2000) and Tzvetan Todorov (1995, 2001). Nora's 'archaeology of the French National Memory' (Hutton 1993), explores French collective heritage and is poised between a history of representations – of the republic, the nation, France – and a history of *mentalités* as, in Nora's words, 'the representative, symbolic and interpretive also have their events, chronology and erudition'. Its focus is not '*what* is remembered but *where* memory work takes place, not tradition but its laboratory' and 'sites of memory' range 'from the most material and concrete such as monuments to the dead and national archives to the most abstract and intellectually constructed such as the notion of lineage, genealogy or even memory-region' (1997b: 15-20). Memory work has been associated with the notion of duty owed to victims of past ills, its excesses or absences – abuses – are being recurrently explored (Nora 1997b, Todorov 1995, Ricoeur 2000); it is depicted as 'saturated' (Robin 2003). The French context lends itself to explorations of the discourses and implications of memory work as national leaders have repeatedly betrayed their concern in its power in relation to sombre episodes of the French past. The shadow of Vichy France loomed large over Mitterand's presidency with revelations and polemics succeeding one another as regards the president's and other prominent politicians' activities during the Vichy era – Mitterrand refusing to condemn the attitude of the

Vichy government (Rousso 1991, 1993a, 1993b, Conan and Rousso 1996). During Chirac's two terms of office, there was an upsurge of memorial laws – on the Armenian genocide, on slavery and on colonisation – yet if Chirac brought some form of closure to the Vichy past with the acknowledgement of the responsibility of the French state in the deportation of the Jews, the page was not turned as regards the colonisation of Northern Africa. In January 2006 a controversial piece of legislation was amended, following intense media debates as regards history and collective memory and a petition signed by historians; it had stipulated that, in the teaching of the French imperial past, emphasis ought to be placed on the benefits of French colonial activities in Northern Africa (Stora 2007). Sarkozy's position on the issue was voiced in his acceptance speech when he rejected the grammar of 'repentance' likened to national self-deprecation. Looking beyond French borders, such highly politicised debates surrounding the manufacturing and use of public memory are shared by many settler societies whose colonial past is at stake in the construction of their modern image and is re-assessed in the light of their present cultural diversity. Emplotting the story of a nation as a smooth flow from beginning to present days is a problematic act which begs, amongst other elements, the question of whose beginning is to be recounted – that of the settlers or that of the original inhabitants; it often leads to a cacophony of fragmented stories on which a communal narrative sits uneasily. In the United States for instance, formerly one-directional nationalist narratives, such as Turner's frontier theory, Manifest Destiny, and the myth of the west, have been transformed and broadened to relate vastly more complex stories (Jameson and Mouat 2006).[1]

In memory studies the capacity for public memory to be used as a tool of social control serving the interests of ruling elites and states, or as a means for oppositional and suppressed voices to express their own vision and challenge dominant interpretations has been equally verified. Emphasising the critical role of the state in moulding collective identity, the path-breaking work of Hobsbawm and Ranger established the notion that the rulers of emerging nation-states 'invented traditions', using symbols and semi-ritualised practices to foster social cohesion, confer legitimacy on institutions and positions of authority, and promote common beliefs and value systems in populations through socialisation. The process ensured the dissemination of official versions of the past emanating from above, serving the current interests of those holding power and eventually validating their legitimacy (Hobsbawm 1983). By contrast, the work of the Popular Memory Group, taking stock of the findings of oral historians, focused on the relations between hegemonic discourses and oppositional or popular forms of expression and on the interaction between public representation and private memory, all affecting the substance of dominant memory (Popular Memory Group 1982). They stressed the primary role of the sense of the past and history in political culture. In the case of oppositional memory, other studies have shown that literary works in totalitarian states often preserve the memory of silenced social groups (Connerton 1989: 15, Winter and

1 I am grateful to Marie Bolton for providing me with this useful reference.

Sivan 1999: 7) while 'unofficial knowledge' shaping people's perception of their past may be gleaned from music, iconography, photography, maps and landscape (Samuel 1996: 3-48). As regards the modern trend for heritage preservation in Britain, Raphael Samuel contradicting the claim made by early analyses which contended that it was the product of a politically conservative élite striving to exhibit a muted, somewhat neutralised, version of the past devoid of pain and conflict and hence unthreatening for the present, demonstrated that the movement in its many-faceted forms was 'not only more democratic than earlier ones but also more feminine and domestic' and 'help[ed] to support both a multi-ethnic vision of the future and a more pluralist one of the past' (Samuel 1996: 161, 308). The issue of authority bound to institutions or agencies sanctioning a version of the past is one which becomes burning when residual, lingering memories are stifled by master narratives giving little scope to sectional views. It therefore comes as no surprise that one prevailing image found in the language of memory studies is that of 'contest', of 'struggle in the terrain of recognition' with groups striving for their version of the past to be acknowledged and represented (Hodgkin and Radstone 2003, Misztal 2003, Ashplant et al. 2004b: 16-32). What is at stake is not the past for its own sake but the past as it reverberates in the present, the past as prey to present priorities and its fluctuating representations are testimony to the imprint of their context of emergence. As Pierre Laborie (2001: 57) has noted, 'memory is less the mark of the presence of the past than the present of the past'. Control over versions of the past are closely entwined with issues of identity and identity construction; individuals and groups in order to locate themselves in relation to others need to gain a sense of their past and their origins; in this process memory is frequently mobilised to serve claims of identity and as a result is prone to manipulation or instrumental use (Ricoeur 2000: 97-8, Lowenthal 1985: 197-200, Fentress and Wickham 1992: xi, 88). 'It is not just that "he who controls the past controls the future" but *he who controls the past controls who we are*' (Middleton and Edwards 1990: 10). What could also be added is that what we are also conditions the manner in which we control the past.

The production and transmission of collective memory

The notion of 'sites' – not exclusively in the geographical sense but as metaphorical inscriptions of the past, as traces with significant mnemonic qualities – consecrated in Nora's work helps to describe the multiplicity of ways in which particular collective memories are supported and transmitted. Typologies and methodologies have been outlined, attempting to pin down the specificity of various forms of collective remembrance and provide a framework for the study of the processes conditioning their emergence and evolution; the following section is a summary of those that have helped mould the thinking behind my own study of Clearance remembering.

Many studies differentiate particular social groups acting as driving forces behind the making of any collective representation, as for instance the work on the collective memory of World War II and the Vichy regime by French scholars (Rousso 1991, Franck 2001). They distinguish between, for example, prisoners of war; *résistants* whose memory is further divided along political allegiances – communist, socialist, or Gaullist; and concentration camps deportees, also further divided between those deported for political or racial reasons. What is brought to light is the tension between the memories fostered by some groups and a 'dominant' or 'official memory' moulded by the authorities whose ultimate aim is to unify those heterogeneous and often centrifugal memories in order to safeguard national unity and identity, with for instance the 'Gaullist resistancialist myth'. Rousso built part of his analysis around the notion of 'carriers' of memory, designating the sources which enable and give shape to the reconstruction of a past event; such carriers are further defined as 'official' such as those orchestrated by national or local governments; 'organisational' when staged by specific associations specially created to preserve the individual memories of group members – soldiers, deportees; 'cultural', namely literary, televisual or film representations; and finally 'scholarly' implying here a factual reconstruction and hermeneutical approach as in the work of historians. However this presentation irons out the complexity of the relations between state and civil bodies and between the various types of discourses formulated.

More recently, also using an event-oriented perspective and drawing from the earlier work of the Popular Memory Group, Ashplant et al. (2004b) have introduced useful analytical tools framing their exploration of the production, dissemination and contest of collective forms of war remembrance.[2] They distinguish between three interconnected factors at work in this social process: the 'narratives' elaborated and their use, rejection or recasting of pre-existing discursive modes and forms; the 'social arenas' in which such narratives are articulated which 'range in social breadth and political importance from the networks of families or kinship groups, through those of communities of geography or interest, to the public sphere of nation-states and trans-national power blocs' (Ashplant et al. 2004b: 17, 34); and finally, the 'agencies' or institutions through which those memories are articulated, and which might be sponsored by the state or be civil in nature, national or local. These structures and conditions constitute key aspects in the process of public recognition of their particular memories – be they dominant or 'sectional' – sought by various groups. Their study also foregrounds the question of generational transmission and insists on the necessity to re-inscribe the individual and the subjective in memory studies looking at the way personal experience is filtered through existing cultural narratives whilst, at the same time, emotion and subjectivity may be mobilised in order to serve political ends.[3] In this respect, if for modern events, eyewitness experience and personal testimony are endowed

2 See also Dawson (2005).

3 Concerning the structure of life-stories, see also Samuel and Thompson (1990: 1-22).

with increasing public recognition as in the case of the Holocaust,[4] as events grow more distant in time, the notions of postmemory and generational transmission are increasingly underscored as an essential part in remembrance processes. First introduced by Marianne Hirsch, postmemory characterises the 'experience of those who grow up dominated by narratives that preceded their birth ... by the stories of the previous generation' (Hirsch 1997: 22). For Ashplant et al. (2004b: 47) the reconstruction of the past from stories instead of unmediated individual experience 'encourages artistic creation' with 'in the absence of direct memory, artistic forms supply[ing] the material of imagination'.

In the case of the Clearances, postmemory is a useful notion to analyse representations of the period in their many-facetted cultural forms throughout the twentieth century – visual arts, novels, plays; in the genesis of many individual reconstructions are found testimonies either transmitted by forebears or part of local tradition and knowledge. Not necessarily transcribed literally, such stories nevertheless acted as powerful stimuli fuelling artistic endeavour; the example of Margaret MacPherson (1972) is a case in point, as not only does her novel, *The Battle of the Braes*, owe much to Skye oral tradition, to stories she heard in her youth from crofters involved in the incidents but also, the reminiscences of her father, Norman Maclean (1953), author of *Former Days*, helped shape her own narrative of the events.[5] Similarly the father of the artist Will Maclean, who has returned many times to the theme of Clearances and their aftermath in his work and whose ancestors hailed from cleared districts, had left a manuscript family history in which he argued against the Clearances sinking into oblivion:

> There are many these days who say that the Clearances should be forgotten and the hatchet buried, but I do not take that view. I think it should be kept alive and every Highlander should have copies of the histories of the Clearances along with the bible in his bookcase. It is not right that the Clearances should ever be forgotten (Macmillan 1992: 14).

Maclean's monumental cairns commemorating the actions of land raiders on Lewis are perhaps the strongest endorsement of his father's view. Incidentally, he was selected by *Cuimhneachain Nan Gaisgeach*, the local group specially formed to promote the cairn project, largely because of this intimacy with the subject (Gouriévidis 2008: 186-88). His Clearances art is at the same time inspired by eye-witness accounts, such as Geikie's depiction of removals on Skye, and resonates with references to later literary works: the poetry of Sorley MacLean and Derick

4 See the work of Annette Wieviorka on the Shoah, in particular Wieviorka (1998), Wievorka and Mouchard (1999). Commenting on the mass of Holocaust testimony Wievorka sees it as 'stand[ing] at the intersection of the individual and society' (Winter and Sivan 1999: 125).

5 Personal correspondence with Margaret MacPherson, 6 August 1991.

Thomson or Neil M. Gunn's novels (MacMillan 1992: 22-25, 52-59, 102). In turn his own creation may generate further narratives.

If a sense of the past is acquired through pre-existing 'narratives' and discursive modes which are recast or re-shaped, this raises the question of the relation between history and other memory-producing media. In memory studies, the relations between historical writing and public memory be it official or oppositional have been thoroughly dissected, the Popular Memory Group (1982: 213) for instance stating that 'political domination involves historical definition [and that] history is a stake in the constant struggle for hegemony'. One approach is to conceive of history and memory in terms of a series of antinomies – a dichotomy which underpins the positions of Halbwachs, Le Goff and Nora (Misztal 2003: 99-108). Another is to regard history as merely one avenue towards acquiring some knowledge or sense of the past: 'history does not have exclusive rights over the production and transmission of knowledge nor over the renewal of a shared understanding of the past' (Valensi 1993: 499). This is the perspective adopted in this book.

History, memory and a 'tense past'[6]

There are many ways in which a sense of the past might be gained and transmitted; history is but one of them. Yet, as a memory medium, in status it might frequently be set apart from others as Rousso's categorisation testifies – 'scholarly' history often acting as a legitimating tool in competing versions of the past and historians thus being invested with colossal social responsibility. Professional historians may be consulted for museum exhibitions, they may also be actively involved in committees organising commemorations and their expertise may be enlisted over issues of historicity of particular public representations of the past or during trials where they stand as expert witnesses.[7] However, as an authenticating instrument, history might offer more ambiguities than straightforward answers as it is prone to dissent and is rarely univocal. History is pluralist both in terms of practice and in terms of genre; historians advance arguments and demonstration predicated on reasoning and deduction: as such their views are potentially open to criticism and refutation.

The practice of history entails selection. Traces of the past come in many forms and historians have to sift through a vast reservoir of evidence – within the scope of the period studied; the history they produce is the result of a succession of choices: their object of study, the sources they analyse, the way they approach and analyse them. They also opt for a specific interpretive system and specific literary devices

6 The expression is borrowed from Antze and Lambek (1996).

7 See Rousso's article (2003) on the Papon trial in October 1997 raising the issue of the dilemma faced by historians used as 'expert witness'. For a wider discussion of the historian's role see Dumoulin (2003).

or genres to organise and construct their narratives. 'The art of historical writing is that of making a master narrative out of chaos' (Samuel 1996: 434). Historians organise, order, label and spotlight specific events, figures, periods. This process of selection might stem from historiographical shifts offering new paradigms and vocabularies: shifts in methodology – emergence of statistical methods or use of oral evidence – or in schools of thought with fresh interpretive and critical stances as with the advent of social or gender history (White 1973, Samuel 1996: 429-44, Pomian 1998). History like collective memory is not a homogeneous entity; it is composed of strata conditioned by historically and geographically specific contexts linked with the pressures of contemporary scholarship and academic practices, discursive modes and models of explanations. It is embedded in traditions – national and scholarly – and practices to which historians seek to adhere or from which they strive to distance themselves and as such can be endlessly revised. It is often this attribute of history, its relativity, which tends to be neglected by memory actors and activists.

One of the recurrent features which appears at first sight to distinguish history from memory is its assumed dispassionate and detached character, refraining from engaging emotions and political or social loyalties, from inviting sympathy and identification. Yet, history-writing as a genre is also conditioned by the public a historian seeks to target and the impact sought; these objectives inevitably affect narrative strategies, style and method. One obvious case in point is popular history, which, with its appeal to a wide audience, in terms of structure, dramatic presentation of information and deliberate emphasis on the emotional sits uneasily with the preconceptions just outlined. Evidently, the creation of proximity is not simply the prerogative of popular historians nor does it simply result from the issue of reception. Oral history has often been singled out for erasing distance whilst social history, by bringing into sharp focus the lives of the voiceless – women, the old, peasants – those whose experiences had been forgotten in narratives concentrating on long-durational macro-history, was able to 'humanize historical writing', some scholars particularly 'attracted by the affective and ideological proximities of the new genre' (Phillips 2004: 128).[8] The issue of distance stands at the heart of historical practice and in this respect history is little different from other memory media more commonly associated with affect.

When addressing events involving pain, suffering and trauma, historians are not only confronted with methodological, philosophical and ideological choices, but they are also affected by moral and ethical considerations. How to deal with the searing words which are unavoidably associated with many of the themes on which they focus: war, epidemics, famines, rebellions, migration and displacement, or strikes? How to translate those experiences without sinking into callous indifference, denial, over-emphasis of the sordidness and pathos of the human condition or even voyeurism? The task of translating private pain is more likely

8 Phillips' analysis encompasses widely more sweeping elements than those mentioned here.

to be the province of writers of fiction. Although historians routinely underscore moments of political and social rupture steeped in mental and physical suffering, adversity is read as the ineluctable consequence of major events or processes and words of pain are seldom analysed and construed. These, however, are not only powerful signifiers of breaks in the social fabric, of a shattered 'imagined community' whose landmarks are being eroded or jeopardised, but also act as triggers for subsequent developments; the experience of suffering disrupts as much as it welds individuals together, giving rise to new modes of thinking and redefining power relations. These might also lead to an assessment of mechanisms with strong bearing on the present, thus highlighting the exemplary potential of history. In Arlette Farge's words (1997: 27): 'By exploring the rationale behind the workings and agencies that spawned [suffering and cruelty], the historian can provide the intellectual means of erasing or avoiding them.' Grappling with such highly charged emotions, becomes all the more crucial as history proposes interpretive bases for the future while participating in the construction of memory. History's and memory's function is to incorporate the past into the heritage of a society, a heritage which, in turn, might mould institutional or mental structures (Martin 2000: 801).

Many social scientists and philosophers have reflected on the function of memory and its media in the context of traumatic events, with which the injunction to remember or never to forget is associated. Two closely intertwined concepts stand at the heart of their reflection: those of justice and debt. Deep unease as regards the manner in which memory is exercised has been voiced; its uses and abuses and the often decisive role of historians and social actors involved in memory work when recovering a tense past are scrutinised. Their analyses, taking stock of what Hobsbawm has described as the 'Age of Extremes', recurrently calls forth two contrasting situations: excess of memory and absence or deficit of memory. One recurring critical issue raised hinges on the notion of victimhood and the use made of the status of victim bestowed on individuals or groups, a status which entails exorbitant privileges and opens up an inexhaustible credit line (Todorov 1995: 28-50, 2001: 3-7, Ricoeur 1998a, 2000: 97-111).

If much of the thinking outlined here surrounds such momentous events as the Holocaust and its aftermath in European countries, some of the theoretical and ethical points put forward are useful to analyse events remembered as traumatic by the social groups and individuals that experienced them. Among the painful processes that Farge (1997: 23) lists as potential active forces in society is the 'discreet and often silent pain of migration, exodus, of people displaced to other regions looking for work far from their traditional affective life'. In a similar vein Samuel and Thompson (1990: 19), in their exploration of the 'myths we live by' – myths not of course used in the sense of falsehood or inaccuracy but in a broader and vastly richer sense, incorporating symbolic and cultural templates resorted to, for instance, in definition of a sense of shared identity – have shown the dynamic power of the sense of injustice and wrong and have used as telling illustration the case of cleared islanders:

> This is why for minorities, for the less powerful, and most of all for the excluded, collective memory and myth are often still more salient: constantly resorted to both in reinforcing a sense of self and also as a source of strategies for survival. In this context it is often persecution and common grievance which define belonging. The Gaelic-speaking crofters of the Scottish Western Isles still recount being driven two centuries ago from their townships by sheep-farming landowners.

Collective memory, articulated around the experience of victimization and loss as much as that of resistance and defiance of rules seen as oppressive, gives rise to a set of values and a world-view defining common identity and purpose. Museums, as public agencies, can potentially give shape to such common identities through the narratives they construct; they re-present a past which divulges as much about the values and priorities of the social group controlling them, as they do about the past itself.

Museums and civil society

Increasingly museums have moved outside the realm of the state and have been developed by an array of social groups. As places where identity can be defined and where dominant narratives can also be disputed, museums are 'integral parts of civil society' (Karp 1992: 5) – the sphere of society acting as linchpin between the citizen and the state. Building on Antonio Gramsci's view of the socio-cultural institutions of civil society as sites where the hegemony of an existing social order is legitimised and perpetuated – as sites for the manufacture of consent – those institutions are now construed as arenas where hegemony can be contested and breached. Museums conveying a sense of the past are one of the terrains of 'contest' in which divergent social groups vie to see their collective memories allocated space and granted recognition.

Much of the discussion regarding museum development and social role centres on the dialectic 'old' and 'new' epitomised by the 'New Museology' whose critical corpus interrogated museum practice and aims and conceived of museums not simply as repositories of objects but as deeply political and ideological institutions because of their legitimising power.

> In the acquisition of material, of whatever kind, let alone in putting that material on public display or making it publicly accessible, museums make certain choices determined by judgements as to value, significance or monetary worth, judgements which may derive in part from the system of values peculiar to the institution itself, but which in a more profound sense are also rooted in our education, our upbringing, our prejudices …

> What is wrong with the "old" museology is that it is too much about museum *methods*, and too little about the purposes of museums (Vergo 1989: 2-3).

The ultimate goal of New Museologists was to transform museums into more democratic and inclusive spaces. Their premise was that traditional practices had made them into little more than instruments for the dissemination of dominant ideologies, usually state-sponsored, legitimating and reinforcing elite authority, social order and values. Their classification and ordering of artefacts buttressed narratives of progress and advance from rudeness to civilisation and were closely entwined with national, imperial, or ideological projects, all too often celebratory in tone. In the museum space, notions of citizenship were strictly limited to encouraging conformity with received normative views rather than eliciting questions (Lumley 1988, Bennett 1995, Huyssen 1995: 13-35, Harrison 2005, Weil 2007).

The unsettling probings of 'New museology' coincided with the rapid expansion of a plethora of museum-type venues from the sisxties and fast-changing socio-economic conditions, which contributed to undermining the former authority of museums as fonts of knowledge and unassailable certainties and to redefining their own image, identity and social function.

> The museum's role as site of an elitist conservation, a bastion of tradition and high culture gave way to the museum as mass medium, as a site of spectacular mise-en-scène and operatic exuberance (Huyssen 1995: 14).

In Britain, the re-structuring of the economy with the decline of the manufacturing and industrial sectors and the growth of services, notably tourism, along with changes in recreational patterns, the extension of holidays to lower income groups and the advent of global tourism acted as catalysts for change in the museum world. In addition, local government reform coupled with the availability of funding from an increasing number of quasi-governmental and other agencies also provided added boost to the opening of museums, increasingly identified as agents of socio-economic development and regeneration. Visitors and communication became key concerns for museums re-orienting themselves as public services in a sector where the boundaries between high and low culture were fast dissolving, the shift cloaked in the mantle of democratisation. Not only did museums attempt to target as broad a spectrum of visitors as possible in terms of age, gender, class, race, nationality but, as in the case of small local museums, they also often involved lay people, the museum world no longer simply the preserve of specialists. The marriage of tourism, concern for the public – or rather the consumer – and culture appeared to turn museums into commodities, part of the 'heritage industry' (Kavanagh 1990: 47, McCrone et al. 1995: 1-26, Fyfe 2006). Reviled by some as the symbol of the decline of England and denoting decay and nostalgic appetite for a glorious past (Hewison 1987, Wright 1985), the heritage industry has equally been identified as a motor for change, re-energising localities. In the words of Barbara Kirshenblatt-Gimblett (1998: 151),

> Heritage and tourism are collaborative industries, heritage converting locations into destinations and tourism making them economically viable as exhibits of themselves …
>
> Dying economies stage their own rebirth as displays of what they once were, sometimes before the body is cold.

From an economic perspective, painful collective memories can be transmuted into valuable assets.

An array of motivations, from the preservation of rapidly vanishing aspects of a shared past to economic factors, therefore cohered in the burgeoning number of museums with their varied and sundry forms, but most of all, the sector saw the involvement of a host of social groups brought together by a sense of common purpose – a concern for their own heritage and its future and, in practice, the desire to create their own local venue or preserve a site to represent it. They are fitting examples of what has been described as 'new sociations', which people join because of

> the emotional satisfaction that they derive from the pursuit of common goals …
>
> [They] provide important sites through which new kinds of identity can be experimented with. They may empower people, they provide relatively safe sites for identity-testing, and they can provide a context for the learning of new skills (Urry 1996: 59).[9]

Whatever the size or location of museum spaces, they are sites where identity is in travail; they project the identity of the group they serve according to shared interpretive frameworks and discursive references stemming from common cultural and historical positions. They may also help shape the view of visitors, yet those visitors come to museums equipped with their own referential systems and socially acquired values and beliefs. Their perception and understanding of the past are equally moulded by a stock of collective memories and assumptions sedimented over time. 'How the present and the past is understood, and how objects are interpreted, depends on personal biography, cultural background, and interpretive community' (Hooper-Greenhill 2000: 119). If the interpretation offered to them is not consistent with their own experience and context, it is likely to be misconstrued, questioned, resisted or challenged. 'Contact zones' is the expression used by James Clifford to describe museums and the cultural encounters or frictions they engender; borrowed from Mary Louise Pratt's *Imperial Eyes: Travel*

9 'New sociations' is borrowed from Hetherington (1994).

and Translation in which it suggested the 'intersect[ing] trajectories' [of] 'subjects previously separated by geographic and historical disjunctures', Clifford expands the notion to 'relations within the same state, region, city' and gives it a social dimension (Clifford 1999: 438, 445). 'Contact zone' may also intimate conflict.

The examples and references used in the literature delineating the development of the critical apparatus and the thinking concerning the social function of museum, their impact on identity formation, and the role of the past in such processes are as enlightening as the theories and professional prescriptions they spawned. The critical role of history-telling in museums – be they small-scale or national – in a world which saw, in the twentieth century, two world wars, decolonisation and the resurgence of various strands of nationalism, the collapse of apartheid in South Africa, the gradual move to dialogue and hesitant power-sharing in Northern Ireland or the implosion of Yugoslavia, is demonstrated through numerous instances emphasising the power of collective memories in making sense of the present and shaping the future. Innovative exhibitions may just as well come to grief on misunderstandings revealing of deep socio-cultural fissures, enduring traumas or simply strained relations, as they may become useful dialogic spaces depending on the context in which they are viewed, the discursive issues they choose to address and the way in which they choose to address them.[10] An often-quoted article is that of Jeanne Cannizzo, guest curator of the exhibition 'Into the Heart of Africa', held at the Royal Ontario Museum in 1989-1990, in which she describes how her attempt at using irony, juxtaposition and muffling the more traditional didactic curatorial tone in an approach which was deliberately reflexive, eventually misfired. The exhibition hoped to critically engage not only with the process and ideological bases of European colonisation of Africa, but also with past collecting and displaying practices, whilst also presenting the richness and diversity of African culture (Cannizzo 1991, Schildkrout 2004). Received as thought-provoking by some, it was interpreted by others as 'racist' disclosing 'a white supremacist view of Africa', a perception which a number of cultural critics partly assigned to recent political developments, namely the tense relations between the African-Canadian community and the police following the shooting of a black youth. For Cannizzo (1991: 157-58), this confirmed the extent to which 'if the museum itself is a cultural artefact, as with individual objects, the context in which it is viewed will be influential'.

In Scotland, this was amply demonstrated when the Museum of Scotland (MoS) was first opened on St Andrews' Day 1998 and it is the relationship with England which casts a long shadow over the interpretation and perception of Scottish heritage:

> As regards the Scottish dimension of heritage, it is obvious that it is doubly peripheral to the dominant but often implicit English one. It is plainly not

10 For specific examples see *International Journal of Heritage Studies*, Karp et al. (1992) and Watson (2007b).

> English, but also escapes hierarchical ordering in the way the north of England ("the North") does not. It has the capacity to generate an alternative political and cultural consciousness (McCrone et al. 1995: 24-25).

The emphasis in Scotland on distinctiveness and an alternative vision finds its most forceful expression early on in the exhibition when visitors are greeted by the Declaration of Arbroath with its patriotic ring. Here the past is harnessed to project a strong and vibrant national identity. Magnus Linklater (1998) pressed the point home in one of the Scottish Sunday papers:

> By drawing together the evidence of a rich and urbane culture with strong links in Europe, reaching back to medieval times, [this museum] challenges head-on the notion that Scotland only began to flourish properly after 1707. Historians and politicians who argue that it was the Union Treaty with England that gave Scotland the means to develop economically and culturally, may have some rethinking to do.

By contrast, *The Economist*, also focusing on the floors devoted to the post-Union period saw in the displays 'much to set the British heart beating … The museum is indeed anything but a one-sided verdict of Scottish history' (*The Economist* 28 November 1998: 42). 1998 was of course also the year when the Scotland Act was enacted amidst fears that devolution might unleash a tide of nationalist fervour. Ironically, posterity might link the advent of devolution with the strong mark of cultural emancipation symbolised by the museum's opening, although this had followed decades of painful gestation and delicate political manoeuvring.

Many of the debates which have framed exhibitions, their making and reception have focused on issues of representation and construction of citizenship, belonging and culture in the context of national, postcolonial and multicultural discourses; they have, not infrequently, centred on national museums or national collections and the tension between the manufacture of master narratives and the representation of socio-cultural diversity, with the plural and often dissonant voices that this implies. A compelling instance of this comes from Australia where the exhibition of its National Museum in Canberra fuelled heated arguments as to its nature and purpose at the turn of the twenty-first century. Critics took a dim view of its deliberate avoidance of a 'whitewash' or the elaboration of a 'coherent' national narrative celebrating discovery, national construction and cohesion, as the museum was designed to act as a space where various stories, voices and viewpoints could be heard – not least that of Aboriginal Australians (Casey 2001, Davidson 2001, Macintyre and Clark 2003, Attwood 2004, Carroll 2004). Its approach was particularly anathema to neo-conservatives – and PM John Howard in particular – for whom the discourse of mutli-culturalism was recast as consensus. While, in Britain, it appears that discursive issues are largely collapsed into the portmanteau notion of 'the people', in post-colonial nations, 'culture wars' have mostly hinged on multi-cultural discourses (Bennett 1995: 109-27, L. Smith 2006: 195-236).

Interestingly, in Australia as elsewhere, it seems comparatively easier in practice – yet no less contentious – to engage with various 'interpretive communities' and promote understanding through more local, community-focused exhibitions in museums with a specific brief; the Migration Museum opened in 1986 in Adelaide (South Australia), – whose philosophy to present a culturally diverse society reflects the multicultural policies formulated in the 1980s – has a community access programme which changes every three months with various community groups producing displays exhibited in a dedicated space, the Forum gallery. Alongside the voice of the 'museum' in the permanent exhibition, the Forum provides a place for distinctly authored stories, themselves revealing of priorities and self-perceptions with which the museum has to negotiate (Szekeres 1992).[11] 'Community', although a multi-layered and fluid notion, has now become a central tenet conditioning museum funding and practice, this social function underscored when linked with such objectives as building or strengthening social cohesion, combating exclusion or promoting dialogue and understanding within and between 'communities'. In Northern Ireland where memories of The Troubles are still being worked through and are strongly divisive, their representation in museums enlisting the participation of community groups offers 'the potential positive impact of working together on interpreting the conflict' (Crooke 2007: 303) as well as furthers general understanding of the role of collective representations in societies grappling with the aftermath of conflict. However, the inclusive ideal, which underpins the 'community' approach, also requires careful navigation if it is to avoid foundering on the rocks of ingrained parochialism and acrimony.

Museums are still implicated in the business of classifying and making order out of chaos, but responding to calls for greater inclusiveness and accessibility, for the representation of more encompassing and less alienating values and narratives, they have had to scrutinise and adapt their curatorial practices and methods. Among the many curatorial issues which have recurrently been explored and surface in the argument of *The Dynamics of Heritage* is that of narrative, more specifically its construction, which, in the museum context, includes: the way in which different media – artefacts, texts, illustrations, photographs, wax models – are combined, spatially arranged and brought to bear on one another; its structure – time-scheme, thematic organisation and approach such as narrow or broad angle views; its symbolic content such as use of dominating tropes; and most of all, its control or point of view. As the museological examples cited above clearly bring to light, the issue of narrative voice is of paramount importance. One effect of the reflection undertaken has been to moderate the 'traditional' use of the 'omniscient narrator', an authoritative institutional voice controlling interpretation and giving little leeway to alternative viewpoints and to questioning (Weil 2007: 42). Another closely related concern has been the use and interpretation of artefacts; in her contribution to Peter Vergo's *The New Museology*, Ludmilla Jordanova (1989: 40)

11 I am grateful to Viv Szekeres, Director of the Migration Museum from 1987 to 2008, for making a copy of her paper and many other sources available to me.

discusses the link between objects and the acquisition of knowledge and concludes that:

> it is essential to understand the grip which the illusion that knowledge springs directly from displayed objects has upon us ... We must lose our childish awe of "treasures" and "wonderful things" in order to replace it with measured appreciation of the awkwardness, the limitations, the downright intractability of objects that, for whatever reason, we endow with value.

This lays emphasis on the importance of interpretation and contextualisation or more to the point the pitfalls of absence of interpretation which 'has come to be seen as elitist and anti-democratic' (Witcomb 2003: 86). Yet as Andrea Witcomb and others have demonstrated, objects are polysemic and contain multiple layers of information and meaning which different 'interpretive communities' may or may not see or choose to tease out (Joly 1998: 74-76, Kavanagh 1990: 107-14). Witcomb illustrates how her own curatorial culture interested in the social dimension of objects had to be married with that of the Portuguese-Australian community with whom she worked and for whom the objects chosen as exhibits supported very personal and concrete memories. Social conditions and political contexts were not necessarily easily mediated through personal experience, and certainly what an object is made to say and the layers of significance it is made to uncover often requires striking a balance between the personal, local and particular, and wider historical trends and processes – a point valid for the way narrative is approached in general. The nature of the museum where history is told also has a bearing – wide-ranging time or geographical brief impose constraints on the degree of precision or conciseness with which a narrative is conveyed. Local museums may well have more scope to delve into the complexity of historical issues, whilst weaving in the local and personal angles, than museums with a national remit. Conversely, privileging the specific and detailed may obliterate wider trends and pressures. These are among the curatorial and methodological issues which will be raised in this book.

* * * *

The Dynamics of Heritage attempts to connect specific museum interpretations of the Clearances with the overall role of museums and museum development in civil society, and considers the texture of museum narratives of the Clearances within the structural, economic and political context of museum expansion in Scotland.

The next chapter focuses on some of the pre-existing narratives and discursive modes which have acted as so many alluvial deposits and as a rich source for museum representations of the Clearances. History will be one such 'sediment', alongside fiction and theatre. The debates at the heart of Clearances historiography will be examined as will the use of history writing as an instrument for ideological and political use. The chapter will underscore the recurrent appropriation of the

Clearances as a symbol and a process in a variety of discourses evolving according to needs, circumstances and context.

Chapter 3 concentrates on the origins of the 'agencies' legitimising collective memories of the Clearances. Given the fact that, until very recently, such representations were mainly found in small museums, the legislation and resources sustaining museum emergence are brought to light. The role of local authorities in the opening of local museums and the nature of the 'sociations' involved with their specific motivations will help centre the discussion on the relative weight of voluntarism and self-help as compared to state input.

The following two chapters move to an analysis of the nature and characteristics of Clearance narratives in museums, Chapter 4 establishing a broad typology with respect to thematic content and looking at the incorporation of pre-existing narratives. In what way local and national museums negotiate between the specific and the general given the atomised and disparate nature of the Clearances? Chapter 5 looks into curatorial practices and methods and the way these have impacted on narratives. For instance it asks whether representations are reflexive and highlight the dilemmas faced in the re-telling of such a disputed subject. Do historiographical debates transpire in museum interpretations? Are clearances presented as a contested issue? Further questions will revolve on how the issue of voices has been addressed, if indeed it has been addressed; and the extent to which the trope of the victim resonates in displays.

The remainder of the book aims at furthering the analysis by connecting the narrative choices made with wider contextual issues linked with museological, economic or political elements. Chapter 6 links features uncovered in the narratives of the Clearances with a historical overview of museological developments and practices looking at the nature and type of exhibitions where the Clearances appear. It seeks to examine the extent to which changes in museum philosophy and methods have affected the ways and means used to convey the Clearances as a narrative and a historical process. Among the points raised are the role of folk museums and the way in which they attempt to address social issues; the extent to which the commoditisation of heritage has compromised Clearances stories; and how the 'social turn' witnessed in museums and resulting in the de-centring of narratives has inflected those of the Clearances. The economic dimension of museum creation will be the subject of Chapter 7; if economic development is one of the motivations stated by some of the social groups involved in the founding of museums, the policies adhered to by funding agencies, in particular foregrounding community regeneration, will be examined. The issue raised is whether Clearance narratives have been modulated by the need to appeal to specific target groups. Finally, Chapter 8 will explore the politics of Clearance narratives. One of the aspects underlined as regards the accessibility of exhibitions has been their relevance to visitors to whom the past is frequently presented as a world that no longer is, a self-contained time capsule with little influence on and connection with the present.

> Past, present and future are often treated in interpretation as disconnected periods and not part of a continuum subject to ongoing processes, causes and consequences (Uzzell 1998: 15).

The extent to which this applies to Clearance narratives will be examined, particularly in relation to land reform and occupation, along with the viability of crofting. More widely, the notion of 'community' will be applied to the different contexts in which Clearance narratives appear showing how the past is used as a pathway to voice the present concerns of different social groups.

Chapter 2
The Legacy of the Clearances

The Highland Clearances is one of the sorest, most painful, themes in modern Scottish history. The events have now receded into the distant past, beyond the direct memory of any living person or even their parents. But the passionate indignation lives on, swollen rather than weakened by the passage of time. A rage against past iniquities has been maintained, fed by popular historians and every variety of media construction. A line of denunciation flows from the oral tradition of the early nineteenth century – the samizdat of an oppressed and angry people – to the electronic graffiti of the present day in the webs of retrospective indignation (Richards 2000: 3).

Eric Richards' words convey the long-lasting legacy of the Highland Clearances, one which modern accounts of the events, irrespective of their nature, repeatedly underscore, at times with exasperation. The small crofters of the Highlands and Islands rank among the archetypal victims of the great social and economic upheaval of eighteenth- and nineteenth-century Scotland, as do the handloom weavers in the urban west (Smout 1987: 393-402). However, whilst the handloom weavers are seen as the casualties of fast-changing circumstances, in the case of the Clearances, notions of injustice and moral responsibility underpin interpretations and pepper contemporary and modern discourses of the events, often pitting social groups against one another. One compelling illustration was, in the spring of 1998, the idea of an apology for the Clearances by Scotland's landowners, relayed in the Scottish press; this, significantly was set against the backdrop of discussions about the pressing need for land reform measures in Scotland. That the manoeuvre reeked of political opportunism and ulterior motives was clear enough (Ogilvy 1998, *Herald* 1998),[1] however, although the Clearances were but an instrumental tool, the intention paraded a sense of collective responsibility and implicitly granted evictees the status of 'victims'; it was testimony to the enduring symbolic power of the events and its significance in the land issue in Scotland.

Yet the equation between cleared inhabitants and victims was not a foregone conclusion in the late eighteenth and most of the nineteenth century in Scotland, let alone in the Highlands and Islands. The emotive, politically- and symbolically-charged character of the Clearances makes their representation in museums a challenging task precisely because of their contested nature and also because museum narratives are subject to constraints of space, and are essentially designed to offer crisp vignettes of the past, limiting in-depth analysis of complex issues.

1 See also Callander (1998) in the same issue of *Scotland on Sunday*, and the letters addressed to the editor in the following issue, 3 May 1998: 16.

The interpretations found in museums are framed by a succession of narratives constructed, re-shaped and altered as time elapsed and contexts changed.

Early discursive formats of the Clearances

In the nineteenth century, as the Clearances unfolded and the process became mired in social crisis and economic failure, eventually leading the government to pass legislation in 1886, various discourses were elaborated, frequently using distinct referential codes – expressions of divergent world views and interpretive or even ideological systems.

In 1985, Richards, whilst recording the changing circumstances of the region and mood of those sponsoring or experiencing its transformation, grouped the evidence that helped shape perceptions of the Clearances around three chronological landmarks. His analysis clustered first around the record of those close to the 'proprietorial order' mainly from the late eighteenth and early years of the nineteenth centuries. Then the more acrimonious debate, which followed as the deterioration of the regional economy was brought to light, opposing apologists of the policies to critics that became increasingly radicalised as famine and destitution gripped the region and class arguments and politics developed nationally. The final milestone was the Napier commission and report along with concurrent works marking the 'historicist' turn – broadly the move from a 'utilitarian mode of thought' to the rehabilitation of the collective and the customary (Richards 1985: 3-103).[2] The 'historicist' vision helped shape the 1886 Crofters Act which curtailed the power of landowners by granting security of tenure to crofters resident in the crofting parishes the legislation defined; the underlying principle which distinguished those parishes was the belief in common use of land (Lynch 2001: 375). Focusing on press records of the potato famine years, Krisztina Fenyo (2000) carves Lowland 'public opinion' of the Highlands and their inhabitants, up into three attitudes: 'contempt, sympathy and romance'. Underlying the chronology and perceptions outlined by both authors are dominant discourses – often complementary – providing the filters through which the Highland situation and the local population were apprehended and recast depending on the context (Womack 1989).[3]

In Enlightenment Scotland, the most pervasive and influential discourse was that of Improvement which formed the national warp through which the weft of local changes was woven.[4] The doctrine of Improvement was much more than the promotion of new forms of cultivation, tenurial organisation and land use and was couched in ways that translated political, social, economic and cultural beliefs.

2 The 'historicist' argument is borrowed from Dewey (1974).

3 This and the following section use some of the argument developed in Gouriévidis (2000b).

4 For a study of early Improvers' schemes, see Youngson (1973).

In the pages of the *Statistical Account of Scotland (O.S.A.)* – a twenty-volume survey of Scotland spearheaded by the improvement luminary and landowner Sir John Sinclair – Improvement, when applied to the Highlands, was premised on the necessity of abolishing traditional practices considered to be hampering agricultural progress, and the twin assumptions that, as a result, prosperity and welfare would replace endemic poverty, hardship and poor returns (*Statistical Account of Scotland 1791-1799*: vol. xx, 340, J.A. Knox 1785, J. Anderson 1785). It took its cue from the tenets of contemporary economists such as Adam Smith along with utilitarian moral philosophy, and was also infused with the language of patriotism. However, patriotism was a fluid notion, encompassing benefits brought to people – tenants and landlords – land, region and nation at large, and smacked of commercial opportunism (Smout 1989: 15). Improvement became associated with the development of extensive sheepfarming presented as a more rational – and lucrative – form of land use, of crofting seen as a novel solution to increasing population, and of alternative employment such as kelping, fishing and weaving. All improvers, in those early days, shared the belief that the Highlands could be transformed and their economy diversified, but not all were favourable to clearances, for instance George Dempster, the owner of Skibo estate in south-east Sutherland. For others, by contrast, Improvement provided the rational justification for removals, particularly when their implementation was marred with allegations of violence and crime and met with public opprobrium. James Loch undertook the writing of his *Account of the Improvements on the Estate of the Marquess of Stafford* to clear the name of the Sutherland family following the contentious Strathnaver evictions of 1814 that he and Patrick Sellar – a name which was to rank high in Clearances demonology – orchestrated (Grimble 1962, 1976, Richards 1999, Richardson 1999, G.M. Fraser 2005, Gouriévidis 2005).

In the discourse of Improvement, the demographics of the region were an essential aspect. Changing contexts, current orthodoxies and expediency impinged on arguments regarding population growth, emigration – forced or voluntary – the consequences of re-distribution of settlements etc., validated by or challenging Malthusian theories. The early phase of emigration from the region were opposed by many landlords who lobbied for the 1803 Passenger Vessel Act – which rendered emigration prohibitively expensive – by calling attention to the regional and national effects of a 'people's clearance' (Bumsted 1982, Richards 1985: 201-26), an emigration chosen by the people to escape rising rents and changing conditions. The departure of this convenient labour force was seen as sapping the vitality of the country and the wealth of estates. However, when, after the Napoleonic wars, the industrial props sustaining employment collapsed and, eventually, in the forties the potato famine blighted the region, Malthusian theories were invoked and emigration was seen as a solution. Malthus' thinking underlay much of the discussions surrounding famine relief and McNeill's *Report to the Board of Supervision of Scotland* – in which he argued in favour of assisted emigration and which led to the 1851 Emigration Advances Act (Devine 1988:

124-25). These views were recurrently opposed by Rev. MacLauchlan (1849)[5] in the pages of *Witness* in which the policies long followed in the region were accused of leading to the *Depopulation* of the Highlands (Gouriévidis 2003-2004). The demographics of the clearing process is an issue which does not lend itself to easy and de-contextualised generalisation, but, as will be seen in the following chapters, it is one on which museum narratives often converge.

Proponents of Improvement often laced their arguments with another discourse which had long coloured representations of the region and its inhabitants: that of difference or otherness implicating inferiority to the Lowlands and their people (Dawson 1998). The logic of Improvement often combined an economic with a cultural strand and schemes had a 'civilising' quality about them. They aimed at reforming the inhabitants' character and morals, in particular what was seen as their fundamental idleness, instilling in them a spirit of industry and self-reliance through, for instance, the introduction of manufacture and a new agricultural system (Loch 1820: 59-60). In the end, it was the entire Highland way of life which was deemed in need of transformation. Partly bolstered by contemporary Teutonist theories, which constructed an 'imagined community' embracing Lowland Scots and their English neighbours and excluding the Highland Celts (Kidd 1995), the discourse of otherness easily sank into rabid racialism. It was most palpable during the potato famine crisis when debates over the causes of Highland distress and issues of relief were tainted by a stereotypical image of the Celts as constitutionally lazy, if not degenerate; within this racial-deterministic framework, the plight of the Highlanders was assigned to abiding intrinsic attributes rather than external conditions (Gouriévidis 2001, 2003-2004). It was voiced in one of its most extreme forms in the page of the *Scotsman* by James Bruce:

> It is now the fact that morally and intellectually the Highlanders are an inferior race to the Lowland Saxon – and that before they can in a civilised age be put in a condition to provide for themselves and not to be throwing themselves on the charity of the hard-working Lowlander, the race must be improved by a Lowland intermixture ... (Bruce 1847: 16-17).

It also impinged on the vision of government officials involved in famine relief (Devine 1988: 125-26).

The shiny side of the discourse of otherness – and one which turned out to be most lucrative in the long term – was the romantic construction of the region, its people and their culture, often forged by outsiders. Highland romanticisation largely stemmed from eighteenth-century aesthetic canons emphasising the sublimity of the landscape, philosophical works recasting the Highlander as 'a noble savage', Ossianic poetry, the literature of Walter Scott, and the Victorian aristocratic cult of tartanry (Withers 1992, Clyde 1995). The work of General Stewart of Garth (1822)

5 His articles were first published in *The Witness* in September 1849 and then in pamphlet form under the title *The Depopulation System of the Highlands*.

is often cited as partaking of this vision. Drawing from his military experience, his *Sketches* do indeed brush a glorious portrait of the Highland warrior and praise the values of the patriarchal society in which he was nurtured; however what moulds his analysis is the racial argument, recast to serve the Celtic race and display the misconceptions and prejudices which underpinned the changes thrust upon it.[6] The *Sketches* contain a condemnation of the introduction of an agricultural system, seen as resulting in misery, pauperism and the destruction of an ancient society, and largely arising from the greed of the local elite. Just as Improvers associated pre-clearance Highlands with backwardness or socio-economic inertia, Stewart of Garth promoted the view of a pre-clearance Golden Age – not the cornucopia of a fertile land flowing with milk and honey, but one where kinship, mutual respect and bonds of loyalty between clan and paternal chiefs prevailed.

Equally significant was the discourse which emanated from Gaelic sources, sympathetic observers or even those who noted its existence only to dismiss it as an anachronistic and unlawful irrelevance. In it, the notion of kinship twinned with that of the tenants' inalienable right to the land occupied for generations – *duthchas* – were crucial and indelible beliefs (Withers 1988: 77-78, 330-32, 414, MacAskill 1999: 24, 32). Gaelic poetry provides insight into the frame of reference and codes which formed part of the imaginaries of the local peasantry from which the poets came – most of them belonging to the crofter and cottar class (Meek 1995: 10). The poems gathered and introduced by Meek (1995) offer the stark antithesis of the ideology of Improvement and rationale behind evictions; to social and material progress, kindness or hospitality and a productive country are opposed loss of home and communities, pauperism, reliance on charity and an impoverished landscape. Clearances for sheep, then deer and grouse moors created a 'wilderness' (MacLachlan 1995: 192, Livingston 1995: 202), but, unlike the romantic canon, it is decoded as one of ruin and desolation, an incongruous and unfamiliar site remembered as cultivated and inhabited. The grammar of patriotism is also inversed as motivations for change are reduced to landlords' greed and their actions to oppression; similarly estate policies are assessed in national terms and the loss of valuable soldiers for the British army explodes the idea of transformation being effected in the nation's best interest:

> I shall say no more about the feats of the brave men
> Who won renown for Britain in every struggle and fray;
> The raising of land-rent is what has thinned our number –
> With no remembrance of Sebastopol and Baclava (J. MacLean 1995a: 206).

The underlying notions of reciprocal loyalty and obligations derived from the clan structure seem transferred to the military sphere. In the earlier poems of the collection, dejection is combined with anger and typically the focus of the condemnation of the Clearances is, by and large, localised, personalised and

6 See his refutation of Pinkerton's view, p. 199.

emblematic, the poems lashing out at specific targets such as shepherds, factors and a few landlords – particularly the new breed of non traditional families (Devine 1994a: 63-83) – rather than the system of landholding or the political structure and legislation upholding it. After 1870, the censure of landholders becomes more systematic and translated into clearer political vocabularies; many of the poems aim at empowering crofters by, for instance, praising land reformers, local land raiders, or drawing attention to the potential of the Napier Commission and the importance of parliamentary representation (M. Macleod 1995: 210, J. MacLean 1995b: 238, Anon. 1995: 240). In the main those later poems were published in the local press and moved beyond the realm of oral transmission displaying the propagandist, but also social power of Gaelic poetry designed to energise the local community, in an age which, nationally, saw the widening of political citizenship. The production of poets helps us to understand the mental structures and cultural assumptions which conditioned the attitudes and collective responses of the peasantry for whom land and security of tenure reigned supreme, and how little compatible they were with dominant doctrines of political economy. This has been interpreted as a clash of ideologies (Smout 1986: 67-68) and, certainly in the later poems, had become crystallised into discernible class polarisation.

From contemporary witnessing to the collective memory of trauma

Not surprisingly, the images, figures and episodes to which socio-cultural and political actors – now and then – tend to return are those that captured newspapers headlines in their days – particularly from the 1840s – and left easily traceable paper trails. In addition to their high profile at the time, their repeated use or reprinting since then have augmented their potency and created a reservoir of powerful accounts and images. The most striking are records of actual evictions stamped with the claimed authority of eye-witnessing – although others denied that the brutalities referred to took place, or at least took place in this manner:

> These proceedings [Farr and Kildonan removals] were carried on with the greatest rapidity as well as with the most reckless cruelty. The cries of the victims, the confusion, the despair and horror painted on the countenances of one party, and the exulting ferocity of the other, beggar all description ... To these scenes I was an eye-witness, and I am ready to substantiate the truth of my statements, not only by my own testimony, but by that of many others who were present at the time (D. MacLeod 1994: 9).[7]

Those penned by Donald MacLeod and Donald Ross (1852, 1853, 1854a and 1854b) were meant to champion the cause of evicted tenants and muster support in their favour at home and, in the case of MacLeod whose work was also published

7 See Withers (2005).

in Canada, within the wider communities of Highland and Island exiles. Whether aimed at attracting charitable contribution at the time of the famine or swaying opinion against clearing schemes, their pieces were not only impassioned and emotional in tone, but also graphically detailed in style, virtually creating a clearance iconography through the scenes they depicted in which the sordid, callous and inhuman were common features. Frequently underscored were particularised individuals and moments of intense crisis, often to the detriment of socio-economic changes apprehended over time;[8] both authors used the power of the spectacle of violence, pain and suffering to appeal to people's compassion and conscience, and question the ethical bases of such proceedings. Hyperbolic and polemic, their accounts were also aimed at refuting or pre-empting counter-representations by providing alternative viewpoints going against the grain of dominant and official discourse. Subsequently, their language and iconography became appropriated in a wide range of narrative genres and contexts and, interestingly, it is from a reprint of MacLeod's work, 'part of or in pursuit of a great deal of left-wing pamphleteering', that John Prebble first learnt of the Clearances in the late thirties (Prebble 1993: 24).

Re-prints or repeated uses of such works over time did much to extend their life in the public domain, and in the nineteenth century none more so than Alexander Mackenzie's *History of the Highland Clearances* (1883), which contributed to the re-inscription and dissemination of previous indictments of the Clearances. Mackenzie's book is a collation of numerous reports of different evictions over time and place, of the Skye agitation in the 1880s and of critiques of the policy and its results. Some of the accounts selected echo the form and style of MacLeod's and Ross' works which are given pride of place – MacLeod's account opening the book; others contain more methodical analyses of the socio-economic foundations of the policy. Mackenzie manufactured a powerful weapon against the landowning and legal systems that enabled evictions, because of the breadth of testimonies he garnered, a weapon skilfully marketed as '*THE history*' *of the Clearances* (my emphasis). In the context of the establishment of the Commission enquiring into crofting conditions, his book was meant to publicise the wrongs of clearing policies and attempt to drum up support for changes in the legislation. Mackenzie hinted as much, in his preface:

> Some people may ask "Why rake up all this iniquity just now?" We answer that the same laws which permitted the cruelties, the inhuman atrocities, described in this book, are still the laws of the country (Mackenzie 1986: viii).

If, in its immediate context, Mackenzie's book was clearly propagandist, as its author sought to promote collective organising, in the long term, it became a memorial to Clearances victims and to some extent to their supporters.

---

8 See in particular Ross (1854).

These sources were only one facet of what constituted the mental construction of the events. In parallel, much of the knowledge and perception of the times would be passed on from word of mouth within and, increasingly, outside the region when, for instance, seasonal workers interacted with fellow Highlanders as they journeyed south and resided away, or as emigrants travelled, established themselves in new countries and formed new communities. Social interaction is an essential component in the transmission of experiential accounts, and the Clearance process itself associated with relocation, temporary migration and emigration but also new economic activities spawned new networks of sociability propitious to the re-telling of past experiences, themselves constitutive of personal identity (Tonkin 1992, Fentress and Wickham 1992). Such milieus – most of them implying a fresh start – inevitably affected the texture of the memory reported. As time went on and as evictions succeeded one another and socio-economic and political conditions changed, personal events were likely to be interpreted in the light of later ones and of changing individual or wider circumstances. Stories bore the imprint of their times and places of re-telling; emigrants' memories recorded in Gaelic songs and poems reflect views of the Old World modulated by New World experiences and show not only many shades of memory not easily collapsed into a single collective strand, but also how memories of the Clearances were toned down or summoned up according to needs, situations and social strata (Bitterman 1999, Kennedy 1999).

Social scientists underline how 'crucial the context of retelling [is] to the nature of memory' (Kirmayer 1996: 178) and how equally crucial is the official provision of a public space for the recovery of the shreds of personal memory, particularly those internalised as traumatic. From this perspective, the Napier Commission is indubitably important '*historiographically* because it gave credence to the popular side of the debate regarding the inhumanity of the evictions and the severity of the social consequences' (Richards 2007: 18, my emphasis), and illustrated the 'historicist' turn of the authorities, but also socially, for the official occasion and platform with which ordinary folks were provided to voice their collective and personal experiences. Although the validity of their vision was not taken as a given, their testimonies and re-telling of past trauma were surrounded with an aura of legitimacy. The final Report itself contains a noteworthy analysis of the value of such accounts and the trappings of memory – personal and collective – one now routinely echoed by many specialists of oral history, and it deserves quoting at length:

> In judging the validity of much of this evidence, we shall do well to remember that these depositions, regarding acts and incidents often obscure and remote, are in many cases delivered by illiterate persons speaking from early memory, or from hearsay, or from popular tradition, fleeting and fallacious sources even when not tinged by ancient regrets and resentments, or by the passion of the hour. But here, in addition … the progress of the Commission was anticipated by agents enlisted in the popular cause … Many of the allegations of oppression

> and suffering with which these pages are painfully loaded would not bear a searching analysis. Under such a scrutiny they would be found erroneous as to time, to place, to persons, to extent, and misconstrued as to intention. It does not follow, however, that because these narratives are incorrect in detail, they are incorrect in colour or in kind (*Report of Her Majesty's Commissioners of Inquiry into the Condition of the Crofters and Cottars in the Highlands and Islands of Scotland* 1884: vol. 1, 2).

The preparation of testimonies became empowering, leading to collective organising. For each area visited, a crofters' and cottars' representative was elected; many read out prepared statements before being examined. Collective and personal commentaries commingled: archetypal details are found alongside more politically-oriented demands. The act of collective remembering is likely to have created a sense of shared experience, reinforcing feelings of belonging and unity, and establishing the Clearances as a moment of social and cultural rupture and as a foundational event. The subsequent development and appropriation of this collective image has been prone to convolutions. For instance, the resurgence of the notion of kinship – particularly amongst expatriate Gaels – including heads of clan, who, often were also evicting landlords, muddied already cloudy waters. Appraisals of the source of common suffering mixed with questions of responsibility and ethics often became fraught with ambiguity and contradiction.

The Clearances and literary imagination[9]

In her study of collective memory, Aleïda Assman singles out literature because it 'disconnects [memories] from individuals, groups and institutions that were once its carriers and reconnects them with an open community of readers' (Assman 2004: 36), which transcends generations, borders and even – when translated – linguistic barriers. It is a point that Iain Crichton Smith made in his fictionalisation of the Strathnaver Clearances, *Consider the Lilies*, in a confrontation between two of his characters, Patrick Sellar and Donald MacLeod, when the latter warns Sellar about the image that Gaelic poetry has created of him, an image that will become fixed for posterity:

> "Patrick Sellar, I see you roasted in Hell like a herring and the oil running over your head …". You see, Mr Sellar, you will become a legend. You have become a legend. Are you flattered? Is that perhaps what you wanted? You talk about the future. Yes, true enough, you too will have a future. Children will sing about you in the streets in different countries, countries you will never visit. They may

9 For more extensive treatments of the literature of the Clearances, see Gouriévidis (1993, 2000a).

> even recite poems about you in the schools. Yes, your name will be on people's lips (I.C. Smith 1981: 144).

Nineteenth century Gaelic verse did contain a host of archetypal images and stereotypical patterns regarding the Clearances. Dehumanising animal imagery was applied to figures on whom anxiety, fear and hatred were projected – Patrick Sellar being the obvious case in point (Baillie 1995: 190-91) and to a lesser extent Sheriff Ivory. Religious symbolism pervaded many poems; evocations of hell fire and devil figures connoted destruction; pre-clearances life was mythologised as a lost Eden and contrasted to the Fall of evictions and subsequent affliction; departure was equated to banishment. In some measure, twentieth-century fiction has perpetuated this tradition.[10] Fire imagery remains a salient feature; factors and sheriffs assume the function of anti-heroes, and the notion of a Golden Age can easily be applied to a pre-clearance era, yet it is not one of pastoral bliss and harmony as few novels tone down endemic periods of poverty and destitution or the tensions and differences inherent to any social group. It is located in the image of a township with its surrounding hills, where cattle could be taken to summer shielings; the security of a house with its nearby plot of land; and the intimacy of a society, where solidarity and communal organisation prevailed. It is a Golden Age in which land and its associated way of life are central.

Most novels are based on actual historical happenings, as rendered in contemporary accounts which are easily identified or transposed into imaginary locations and contexts. Importantly, many are considered as children's fiction and tend to be packed with action and heightened drama. Some tell of the bitter experience of expulsion and rupture. The three best known and most frequently cited as emblematic of the literary treatment of the Clearances are those which dramatise the Strathnaver Clearances and largely contribute to the dominance of the Sutherland Clearances in collective memory: Neil M. Gunn, *Butcher's Broom* (1934), Fionn MacColla, *And the Cock Crew* (1945) and Iain Crichton Smith, *Consider the Lilies* (1968). Reviewers said of Gunn's and MacColla's novels – two of the earliest to be published – that they aroused feelings of indignation, anger and pity at the treatment meted out to Highland people (*Scots Magazine* 1934, *The Scotsman* 1945). Another well-reported eviction is found in Mollie Hunter's *A Pistol in Greenyards* (1965) which draws heavily from Donald Ross' *Massacre of the Rosses*. In her preface, she claims that 'a *true* account of the brutal way in

10 The works included in the discussion are (by order of publication): Ian MacPherson, *Land of our Fathers* 1933; Neil M. Gunn, *Butcher's Broom* 1934; Nigel Tranter, *The Gilded Fleece* 1942; Fionn MacColla, *And the Cock Crew* 1945; Allan Campbell McLean, *Ribbon of Fire* 1962; Kathleen Fidler, *The Desperate Journey* 1964; Mollie Hunter, *A Pistol in Greenyards* 1965; Allan Campbell McLean, *A Sound of Trumpets* 1967; Iain Crichton Smith, *Consider the Lilies* 1968; Margaret MacPherson, *The Battle of the Braes* 1972; John J. Graham, *Shadowed Valley* 1987; David Craig, *King Cameron* 1991; Judith O' Neill, *So Far from Skye* 1992.

which [the clearance of Greenyards] was carried out is given in the early chapters of this book' (my emphasis) and adds that it 'has been written as a salute to the memory of the people of Greenyards, and especially to that of the boys and girls who kept the look-out on Ardgay hill'. Ian MacPherson (*Land of our Fathers*, 1933) and John J. Graham (*Shadowed Valley*, 1987) locate their narratives in areas where Clearances are less widely known but closer to the authors' own geographical origins – Strathspey and Shetland. Uniquely, MacPherson focuses on a landowning family, the Graemes; the father George is a lowland farmer determined to transform his wife's estate; she, as a former Grant of Tulvey, has remained a Highlander at heart. The couple personify a 'cleavage' or 'spiritual divorcement' (MacPherson 1933: 38, 46). Another exception is Nigel Tranter in *The Gilded Fleece* who sentimentalises the issue with a young English factor sent to the Highlands to oversee clearances for sheep, that finds himself empathising with the Highlanders' plight and culture, rousing them to stand up for their livelihood and survival, eventually emigrating to the New World with evictees, and falling in love into the bargain. The social role of spiritual leaders – local ministers and priests – is often interwoven with the experience of eviction and displacement, the moral dilemmas that their position engendered, or their unwavering support for their patrons' policies being pivotal to the plots of such novels as *And the Cock Crew*, *Consider the Lilies* or *The Gilded Fleece*.

Other authors chose to commemorate a 'fighting spirit' and highlight popular defiance and resistance to authorities and Establishment. Both Margaret MacPherson (*The Battle of the Braes*, 1972) and Allan Campbell Maclean (*Ribbon of Fire*, 1962; *A Sound of Trumpets*, 1967) evoke crofters' actions during the Skye Land Wars, whereas David Craig's *King Cameron* (1991) foregrounds protest in Sollas on North Uist during the famine, and Mollie Hunter, the Greenyards confrontation of 1854. Additionally, emigration is frequently part of the plots of Clearances novels. In some cases, departure, epic journeys and settlement in new homelands determine narrative structures as with Kathleen Fidler's *The Desperate Journey* (1964) or Judith O'Neill's *So Far from Skye* (1992) which follow protagonists from Sutherland to the Red River Settlement and Skye to Australia in tales of survival and victory over hardship. In a few novels, emigration simply features as a narrative tool meant to reinforce a sense of anguish, and suggest the crushed spirit of a community. Alternatively, in novels seeking a happy ending, it connotes hope and is given a brief final glimpse:

> It was not a lament that the piper played any more, it was a challenge, MacVarish Mor's Challenge, as the *Charlotte* sailed out into the west (Tranter 1993: 220).[11]

All feature the human drama that disruptive events such as evictions or confrontation with authorities inevitably produced, and bring it to life through

11 See also M. Hunter (1988: 192).

character development inviting empathy or distanciation. Fiction humanising historical episodes invokes subjective identification; novelists take readers into the homes, communities and mental worlds of Highland and Island folks. They make them share their hopes, fears, discords, toil and physical pain; see them grow up and age in increasingly harsher conditions and changing external circumstances; and feel the spiritual, mental and emotional predicaments that those conditions generated. By and large, the most developed and complex characters are those belonging to the world of the peasantry, whilst those epitomising Improvement are frequently reduced to one-dimensional standard-bearers for an ideology and its implementation – little more than stereotypes in some cases. If moral implications trickle into narratives, seldom are they unambiguous when it comes to apportioning blame; factors and ministers do tend to be scapegoated because of their proximity to the peasantry and as perpetrators or defenders of expulsion. The landlord class often seems a remote influence. Additionally, even when economic and political arguments are introduced, they are overshadowed by the localised human impact. However, one unifying aspect runs through novels: the discrepancy between two different mental worlds.

The artists who explored the Clearances had no direct memory of the events, but, with the passage of time, they were able to tap into an increasingly rich seam of Clearances stories and images made available through a variety of publications, not least the popular histories of Ian Grimble's *The Trial of Patrick Sellar* (1962) and John Prebble's *The Highland Clearances* (1963) which revitalised nineteenth-century popular records. The continuing resurgence of the theme of the clearances in fiction has both personal and wider implications. At a personal level, with some, artistic creation was stimulated by the fact that authors had grown up in the shadow of past eviction, migration or emigration undergone by forebears. Time and again, personal experience of place and family ties are brought to the surface. Allan Campbell McLean's great-grandparents originally from Mull, 'had been obliged by economic clearance to exchange their island living for a slum tenement in Glasgow' (Hutchinson 2005); he settled on Skye where his Land War novels are set. There he befriended Margaret MacPherson, to whom *Ribbon of Fire* is dedicated, and whose grandfather was a schoolmaster in Braes at the time of the agitation. John J. Graham 'gathered a lot of general background to the Weisdale Clearances from [his] mother who was born in Weisdale and carried with her much of the oral tradition'.[12] The Australian author, Judith O'Neill, dedicated her novel *So Far from Skye* to her ancestors from the island who emigrated to Victoria with the help of the Highland and Island Emigration Society in 1852. Perhaps even more tellingly is the example of David Craig whose novel, *King Cameron*, is partly dedicated to the people of Sollas; it was published shortly after his journey through cleared districts and emigrant destinations interviewing descendants and piecing together fragments of clearances memory, in particular those of the 'Battle

12 Personal correspondence with J.J. Graham, 10 August 1991.

of Sollas' (Craig 1990: 51-68). Unlike Mollie Hunter, few authors profess 'truth', John Graham (1987: Foreword), echoing Neil Gunn, aimed at

> convey[ing] an impression of life in rural Shetland about the middle of the last century and the shattering effect on one community of a policy of clearances.
>
> The removal ... from Upper Weisdale ... became imprinted in folk memory as a bitter and dramatic blow.[13]

It is not 'truth' linked with historical accuracy, but with human experience and the texture of folk memory, a memory gleaned from personal connections or affinities, and assembled from contemporary or modern accounts.

The moment of re-creation and the thematics foregrounded in the books are also crucial. If in resistance and emigration novels, celebration of past actions and debt to forebears underlie narratives, in others the events recalled involved 'mental complications' (Hart and Pick 1985: 103). Neil M. Gunn underscores the strains that the process of retelling entailed for him; writing the Clearances meant coming to terms with 'the shame of the thing ... because our own people did it' (Hart and Pick 1985: 103). He was born in 1891, grew up a few miles from cleared glens in Sutherland, and recalls that he heard little about the Clearances in his youth both at home and at school; 'presumably our parents did not care to warp growing minds with old wrongs ... Perhaps they [the Clearances] were not only too near in time for school lessons but also too near to the bone' (Gunn 1954). His novel broke the silence of textbooks and classrooms. In his essays, he frequently brings to light his perception of the impact of the Clearances and underlines their psychological cost in Highland people: a 'general lack of belief' or 'defeatism bred of a paralysing history' (Gunn 1987a: 158, 1987b: 163). Significantly, in 1958, one literary critic marked Gunn's and MacColla's novels out as illustrating 'a new attitude towards the "Clearances"' seen 'as the buried root of contemporary Highland difficulties, and as the immediate cause of the crofting problem. History is here regarded as the matrix of the present' (Wittig 1958: 324). The 'Highland problem' was no new concept, but long-standing socio-economic malaise tightened its grip around the region in the late 1920s and 1930s, exacerbated by more widespread economic crisis. The situation of the Highlands became a focus of interest for economists, politicians and intellectuals. Early Clearances novelists addressed an issue which had yet to be studied and analysed in some depth by historians, and literature opened a space for the representation of a past whose effects on the mental and cultural spheres had, thus far, not been fully perceived nor discussed.

For many writers, tackling the themes of the Clearances also has had wider implications of a political cast. Both Gunn and MacColla were writing at a time of

13 Gunn said of *Butcher's Broom* that 'there is little of the Clearances [in it]. The tragedy is the destruction of a way of life, and the book is more about what is destroyed' (Hart and Pick 1985: 103).

resurgent nationalism in Scotland and had joined the National Party of Scotland in the late twenties.[14] If in Gunn's novel, his political views are not forced into the narrative, MacColla's political reading of the Clearances as part and parcel of a long process of assimilation is one of his novel's fundamental components conveyed in long diatribes.[15] Remarkably, MacColla, in 1930, was involved in an aborted plan to entice crofters to organise land raiding on the island of Rum[16] in the hope of attracting the attention of the press and the public to the 'realities of Scottish history'. In his irate autobiography, his avowed objective was '*to let the large world know about the Clearances, that long-continued genocidal episode*', and force the Labour government to take a stand in the Highlands leading to 'the end of the Sporting Estate System ...; the repopulation of the Highlands – and with a Gaelic population ... *In effect, the Clearances in reverse*' (MacColla 1975: 90-95). Not insignificantly, in the 1970s, among the novelists celebrating crofters' struggle to recover lost rights and gain legal protection, Margaret MacPherson and Allan Campbell MacLean, also stand out for their outspoken views on contemporary Highland affairs. Both active Labour Party members, they were vocal and relentless opponents of Labour's policy in favour of transforming crofting tenure into owner-occupation, and were dubbed, along with Brian Wilson, then editor of the *West Highland Free Press*, the '"Highland Luxemburgist" faction' (Ascherson 2002: 173).[17] Their novels in the sixties and seventies arguably spoke to the land issue that was back on the agenda. In some cases, although the Clearances had wider implications for the novelists concerned, their narratives did not necessarily serve as a platform for modern political expression, but were memorials to Highland and Island folks and folkways.

Depending on its form and nature, literature may also have a more overt political use and mobilising power. Alternative theatre in the seventies was such a medium reaching out to audiences in order to galvanise them by promoting political arguments. In the words of John McGrath, one of its foremost exemplars: 'the theatre is by its nature a political forum, or a politicizing medium, rather than a place to experience a rarefied artistic sensibility in an aesthetic void' (McGrath 1981b: 83). There is no better example than his *The Cheviot, the Stag and the Black, Black Oil* to illustrate his use of Clearance narratives, motifs and archetypes, honed to fit a political vision of the Highlands' and Scotland's present and future – at a time when devolution and North Sea Oil were hotly debated Kershaw 1992: 132-167, Gouriévidis 1993: 265-81, DiCenzo 1996). *The Cheviot*

14 On Gunn's significant contribution to nationalist politics see Brand (1978: 102 and 196-227), Finlay (1994: 71-125), Hart and Pick (1985: 108-24).

15 See Chapter 7 where the author's mouthpiece, the poet Fearchar, presents the Clearances as one aspect of the much wider scheme of English imperialism.

16 See Marr (1992: 75-84) and Finlay (1994: 71-125) for further details on Clann Albain with which these plans were associated.

17 See M. MacPherson (1985: 19-21); all three resigned as members of the Labour Party's working group on land reform over the government's Crofting Reform Bill in 1976 (*WHFP* 1976).

glides through 200 years of Highland history, from the Clearances, through the extension of shooting grounds to the impact of the discovery of oil and presents an overtly Marxist analysis of the impact of capitalism on the Highlands. The themes of exploitation – of land and people – and of the arbitrariness of the capitalist ethos ran through a play designed for a popular audience and incorporating working class forms of entertainment whose ingredients McGrath (1981b: 53-60) defined as: directness, comedy, music, emotion, variety, effect, immediacy and localism. The representation of the Clearances alternates between the satire of economic policies and greedy landlords, the pathos of Gaelic songs and the sheer farce of Patrick Sellar's alleged travesty of a trial. Although the Sutherland example provides the play's backbone, local instances of removals were introduced, as the play was performed round a score of local venues in the Highlands and Islands. The author's ultimate aim, however, was to promote assertiveness, to 'break out of the "lament syndrome". For every defeat, we would also celebrate a victory … for every oppression, a way to fight back' (1981a: xxvii-xxviii), epitomised by crofters' actions and success during the Land War. Although for McGrath (1981a: 77): 'socialism, and the planned exploitation of natural resources for the benefit of all humanity, is the alternative the play calls for', its message was open to other readings and appealed to the SNP, whose chairman asked the company to give a performance after the party's annual conference. The issues it addressed – land use, landownership and local interests – and its emphasis on the fact that determinism has no validity in politics and people can act on their destiny, no doubt resonated among nationalist supporters.

What reinforces the power of literary representations of the Clearances is their coherence and consistency over time; as such they contributed to cement the human and psychological perception of the Clearances in collective memory, the texture of which was largely distilled from folk memories.

The instrumental use of the Clearances in political discourse[18]

Just as *The Cheviot* … stirred both nationalists and socialists in the seventies in the context of devolution and socio-economic commotion, so the Clearances have been a recurrent motif in political discourse over the twentieth century – a motif sharpened to fit the needs of a particular agenda and its specific conditions. The Clearances became a powerful vehicle for emotional responses, often enlisted to promote ideas and demand for change, a testimony to their symbolic force within regional, but also national memory (McCrone 1992: 50). The symbolic appropriation of the Clearances over the twentieth century waxed and waned according to needs and circumstances, but it seems to me that three main instrumental uses of the clearances in political discourse can be identified; they are related to land reform

18 Some of the thinking for this section was initially developed in Gouriévidis (1993: 219-323).

and landownership; environment and conservation; and the concept of cultural inferiorisation.

The centrality of the land in Scottish culture and identity has been strongly emphasised by social scientists (McCrone 1997) and land issues, have often occupied centre stage in Scottish political discourse. More specifically debates have erupted relating to questions of landownership patterns, land register and access to the countryside; in the Highlands, these have been twinned, with concerns as regards land use and the conditions and status of crofting. The Clearances have become inextricably linked with issues of landlordism.

> The Highland Clearances have become one of the great scars on the Scottish psyche, and have helped to create a powerful legacy of anti-landlordism which has been a feature of Scottish politics since the 1880s (E.A. Cameron 2001: 88).

The trope of the Clearances which fuelled anti-landlordism was appropriated by both radical Liberals and Labour politicians, some of whom had been leading members of the Crofters' Party, such as Gavin Clark, one of the founders of the Scottish Labour Party (E.A. Cameron 2001: 90-91). A particularly scathing socialist critic of the landowning class and advocate of land reform – or of the House of Lords – was the young Tom Johnston. His *Our Scots Noble Families* (1909) opens with 'a general and comprehensive indictment of Scots landlords as a class', before homing in on prominent dynastic lines and listing evictions, rack-renting and obstruction to the enlargement of crofting land as so many examples of the predatory practices of families who 'snatched the patrimony of the people' (Johnston 1999: xxxi-xxxii). Published a year earlier in article form in the columns of the socialist weekly *Forward*, the book was a bestseller although as Johnston's career progressed and he became part of the political establishment, it turned into a festering sore and was allowed to go out of print. In the same vein, his *History of the Working Classes in Scotland* (1920) was reprinted in 1946; importantly it addressed the Clearances from a national angle, encompassing the whole of rural Scotland, and begins with the 'enclosures' and depopulation of the Lowlands, thus legitimating demands for far-reaching land reform and testifying to the re-centring of the issue nationally – at least at the time of writing.

The variations and modulations of the representations and uses of the clearances in the landowning argument are obviously context-bound. In the first quarter of the twentieth century, some were predicated upon racial discourse, which international politics made increasingly unpalatable and treacherous as a rallying cry after the second world war. Two men, Ruaraidh Erskine of Marr and John Maclean, were influential in framing and using a Celtic political archetype, encapsulated into 'the communism of the clan'. Yet they came from different political horizons and their construction of Celtic communism stemmed from distinct schools of thought. Ruaraidh Erskine of Marr was one of the leaders of the Scots National League, founded in 1920 and 'arguably … was to do more than any other individual to implant Gaelic national ideals in Scotland in the first quarter of the twentieth century' (Finlay 1994: 31, see also Hanham 1969a: 126). His nationalism was founded on the belief in intrinsic racial disparities between Celts and Saxons and, after the Bolshevik revolution, was infused with the ideals and institutions that the Russian experience generated (Erskine of Mar 1918a, 1918b). The two ideas coalesced in his vision of the primitive communism of the Scottish clans destroyed by the 'landocracy' of the Highlands during the Clearances. If the 'communism of the clan' was used as a model to be adopted by an independent Scottish State, Erskine had little to say about the communal organisation of townships or the practice of runrig agriculture and joint-occupancy system. The Celtic bases to which he harked back were more of a romantic than socio-economic cast (Gouriévidis 1993: 248-51).

Romanticism was a far cry from the convictions of the red Clydesider, John Maclean who moved 'in the direction of revolutionary nationalism with its roots in the Scottish past' under Erskine's influence (W. Knox 1984: 188). Initially, his concerns and public addresses seldom focused on the Highlands, but on the central industrial belt. Nevertheless, the Highlands, their history and the landownership system were often referred to, for instance to justify his own socialism: 'when I stand true to my class, the working class, in which I was born, it is because my people were swept out of the Highlands' (Milton 1978: 111). When used as a trope, the Clearances were invoked as a parallel to the situation of the workers in the Clyde valley and elsewhere in Scotland and connote capitalist exploitation and betrayal: '… officials … have betrayed the workers (as Highland chiefs … have betrayed their peoples in the past to the English)' (Milton 1978: 85). Later on, when he endorsed the idea of the 'communism of the clan', the notion of betrayal became related to the Anglicisation of Highland chiefs, corrupted by English values – of which capitalism – and enforcing 'the English landlord policy of land clearances' (Milton 1978: 217). Maclean's crisp use of the Clearances is inserted in Marxist arguments of class exploitation, on which the notion of race is superimposed; his allusions, unadorned with details and very concise, illustrate how potent a symbol the period had become in Scotland.

Another feature of those modulations is the use of the Clearances coupled with demand for land reform also associated with Home Rule and the legislative potential a separate parliament would offer. In Labour rhetoric, the association mirrored the party's flirtatious relations and the oscillating strength of its commitment to the principle. In the 1920s, a pamphlet written by J.M. Macdiarmid (1926: 5, 17), Labour candidate for Kinross and West Perthshire, provides a blistering critique of the landholding system and the British Parliament whose dealings with Scottish affairs led to 'the desolation and economic ruin of the Highlands … the evil work which began with the barbarous clearances of the Highland peasant a hundred years ago'.

Attacks on landownership structures also seeped into the literature of pressure groups, such as those campaigning for the right of access to uncultivated mountains and moorlands for recreation. The campaign was first initiated in 1884 when the Liberal MP for South Aberdeenshire, James Bryce, introduced his Access to

Mountains Bill. Versions of his bill were defeated over a dozen times between 1884 and 1939. Ernest Baker, in one of his early mountaineering reminiscences, denounced the system of landownership:[19]

> There must be something vicious in an economic system the by-product of which is that amazing anachronism, a country of the size of the Highlands stripped of its inhabitants, its natural resources squandered, and the whole turned into a rich man's preserve ... To say that the Highlands are still under feudalism would be unfair to the feudal system (Baker 1923: 45-46).

Landownership was singled out as the root of such aberrations as the Clearances or denial of access to mountains, a parallel was drawn between the contemporary socio-economic predicament of the region and the issue of access.

Most of these examples, from the early twentieth century, confirm the view that land reform has not been a constant of Scottish political debate since the 1880s; it was a prime concern until the 1920s and then returned to prominence in the 1960s (Cameron 2001: 84, Warren 2002). The socio-economic plight of the crofting areas, however, was addressed separately through the establishment of the Taylor Commission, which included amongst its members both N.M. Gunn and Margaret MacPherson, and resulted in the creation of the Crofters Commission in 1955.[20] Whether the lifeline of crofting was interpreted in terms of amalgamation of holdings as was the case in the 1950s (Mather 1985: 221) or, subsequently, in terms of owner-occupation, the spectre of the Clearances was prone to be resurrected (Gouriévidis 1993: 232-36). The question of owner-occupation became an intensely divisive issue in crofting politics, and was given an unexpected twist with the movement of community ownership re-ignited by the Assynt crofters in 1992 (Warren 2002: 48, MacAskill 1999). At a national level land reform seemed to return to the forefront of Labour political agenda when the devolution issue was re-considered and nationalist support crept up in the polls. Over the last quarter of the twentieth century, the fate of crofting and more widely the landowning question eventually achieved increased prominence in the pages of the *West Highland Free Press*, founded in 1972 and edited by Brian Wilson,[21] through the research first undertaken by the socialist John McEwen (1977) in the late sixties charting patterns of landownership in Scotland, and through the

19 Ten years later, when access to mountains relaxed, his tone towards the landowning class was to alter significantly, see Gouriévidis (2004: 142).

20 Margaret MacPherson wrote a note of dissent arguing in favour of state ownership of land in the crofting counties (*Report of the Commission of Enquiry into Crofting Conditions* 1954: 89-93).

21 Brian Wilson (1997), on the 25th anniversary of the radical paper, described it as 'the champion of land reform – the alpha and omega of Highland politics'. James Hunter was given a historical column which started with the centenary of the Battle of the Braes in 1982.

various activities and copious production of James Hunter. Since the late 1970s, Hunter has been involved in numerous pressure groups and quangos – and lately the University of the Highlands and Islands; differing though their specific aims were, these organisations cohered around issues of regeneration, repopulation and socio-economic and cultural revitalisation of the Highlands and Islands. In his writings, the Clearances are a potent leitmotiv, an engine which drives many of his analyses of the roots of the region's modern ills. These analyses are predicated upon a historical perspective, first developed in a PhD published in 1976, *The Making of the Crofting Community*, which established his reputation. His aim was to 'write the modern history of the Gaelic Highlands from the crofting community's point of view' (Hunter 1976: 5), and rehabilitate the peasantry's perception of the Clearances, using a Marxist framework of analysis foregrounding class conflict. During his involvement with Highland affairs, he voiced his support for community ownership, a vision which he advocated when he was Director of the Scottish Crofters Union (SCU), and which, then, in the main ran against the grain of crofting opinion (Gouriévidis 1993: 310-14); it was to move centre stage in the 1990s.

The Clearances have also been used as a powerful illustration of the harmful impact on the land of human policies, in environmental and conservation discourse. If, within a century of the introduction of sheepfarming, articles appeared in the *Transactions of the Royal Highland and Agricultural Society of Scotland* – the proprietors' organ – on the deterioration of the soil, its fertility leached out through intensive sheep-grazing, no critique was expressed of correlated 'removals' (J. Macdonald 1877: 193-94, 205, 1880: 82-86, Latham 1883). In the mid-1920s, the deterioration postulate was further developed by environmentalists such as F.F. Darling (Smout 1993) and the Clearances were reassessed as a turning point in the history of Highland ecology, a period when agricultural and social changes intensified and the eco-system was drastically modified. The deterioration postulate was emphatically argued in both socio-cultural and ecological terms by F.F. Darling (Richards 1985: 119-24). Although his findings impressed the Taylor Commission, little was done in practice (Smout 1993: 21).

By the seventies and eighties, environmental issues and 'green consciousness' were no longer fringe concerns; they reflected societal changes and were also given increased legitimacy through European and global pressures and legislation (Warren 2002: 180-83). Environmental concerns also came to embrace conservation and natural heritage, priorities which, at first, were not easily reconciled with arguments in favour of the development of crofting, resettlement and land use (Gouriévidis 1993: 239-43, 308-14). However, the ever-increasing profile, political influence and financial clout of environmental and conservation lobbies meant that their views could no longer be ignored and indeed could be harnessed for the benefit of crofting. Environmental discourse progressively filtered into the pronouncements of many organisations, not least the Crofters Commission which argued that:

> It is essential for social, economic and ecological reasons that the quality of the crofting environment should be maintained and enhanced. There should be

> formal recognition of the contribution that crofting can make to conservation and the environment and encouragement and assistance should be available to crofters accordingly (Crofters Commission 1991: 21).

The environmental reading of the Clearances appears in the literature of a variety of bodies and pressure groups of the period. Drawing on F.F. Darling's thesis of the Highlands as 'a wet desert', the Highland Green Party, in 1989, issued a Rural Manifesto which drew a picture of the future of the region built around the principles of the forest economy and community stewardship of land. In it the Clearances were equated to a destructive force:

> What happened at the end of [the 18th century] tore the heart out of the Highlands as it tore the people from the land and sounded the death knell for a way of life, a way of husbandry, indeed a whole ecology (Land Use Working Group of the Highland Green Party 1989: 4).

At about the same time, James Hunter, then Director of the SCU, underscored the need for crofters to join forces with conservation groups to ensure the future of crofting (J. Hunter 1989: 8). Subsequently, the SCU, in 1992, published a report in collaboration with the Royal Society for the Protection of Birds entitled: *Crofting and the Environment: A New Approach*. In its argument, a parallel was drawn between the threatened croncrake and crofting, and history was used to vindicate the value of crofting agriculture. Notably, a paragraph seeking to demystify the 'wilderness fallacy' made the case for a reversal of the Clearances process:

> Many of the most unpopulated localities were once thickly peopled – their populations having been expelled forcibly by the landlords responsible for the Highland Clearances. Crofters still aspire to win back these lost lands. There is no environmental reason why a measure of resettlement should not occur. Indeed there would be every reason to welcome reforms which made it possible to combine an element of resettlement with, for example, woodland regeneration and other measures designed to rehabilitate habitats which have been seriously degraded in the period since the Clearances (Scottish Crofters Union and Royal Society for the Protection of Birds, 1992: 15).[22]

The repopulation of deserted glens twinned with environmental and cultural protection also form the backbone of the argument of James Hunter's *On the Other Side of Sorrow* (1995).

An important aspect of Hunter's contention in this book is his appropriation of Frantz Fanon's analysis of the economic and mental ravages of colonialism on the colonised, in particular the notion of cultural inferiorisation leading to

22 Incidentally James Hunter, who by then no longer was Director of the SCU, helped write this document.

self-deprecation and loss of confidence, a process which he co-opts for the Highland situation (J. Hunter 1995: 35-37). Long before the Highland condition was theorised in those terms, the Clearances had been marked out, by cultural actors, as the climax, the final crushing blow of a long war of attrition waged on Highland culture. The 'defeatism' inherited from history which Gunn underlined in his essays was a point he repeatedly raised. When he toured the Highlands with the Taylor Commission, it was made to account for the crofters' defensive attitude towards the Crofters' Act of 1886 and any attempt at tempering with the bases it established.

> As a member of the Commission I confess I was surprised at the somewhat cool or guarded reception we sometimes received, for I knew how earnestly the whole Commission was bent on discovering ways and means of regeneration.
>
> Then one night at a public meeting in the Outer Hebrides two or three elderly crofters revealed the hidden objections by telling us straight that they did not want any Commission visiting them or having anything to do with their crofting system. In a flash I understood: they were afraid that we might interfere with the one wonderful marvellous happening in the last two or three centuries – namely the Crofters Act of 1886 which gave them security of tenure (Gunn 1966: 6).

Gunn's 'sensitive and understanding approach to the whole problem' was 'reflected in the Report' (Taylor 1954), in which, from the outset, it was stressed, that 'the history of the past remains vivid in the minds of the people and, in some measure, conditions their attitude to current problems' (*Report of the Commission of Enquiry into Crofting Conditions* 1954: 10). Any changes proposed in the crofting system had to be mindful of these sentiments and the initial aim of the Crofters Commission, established as a result of the Taylor Report, was to restore the crofters' faith in himself (*Report of the Commission of Enquiry into Crofting Conditions* 1954: 36, *Clarion of Skye* 1956: 1).

Later on, James Hunter, by then Board Member of Highlands and Islands Enterprise, used the same argument to explain the lack of enthusiasm shown by crofters to proposals of community ownership. His analysis of their rejection of the transfer of government-owned estates to crofting trusts was due to feelings which

> … lie deep in our history; in the extent to which the quite deliberate destruction of community and culture, right across the northern half of Scotland, undermined our self-belief, self-confidence, in ways from which we suffer still (J. Hunter 1992: 6).

This goes a long way towards explaining Hunter's tone, when, as chairman of HIE in the late 1990s, he was keen, in their yearly reports, to systematically underline 'success stories' and the achievements of the Network, the region and its inhabitants not only in socio-economic, but also cultural terms, laying emphasis on a restored sense of confidence and renewed pride in local heritage – a heritage much maligned in the past. From this perspective, encouraging community ownership meant empowering local groups and building self-esteem (HIE *Annual Report 1998-99*: iii, *Annual Report 1999-2000*: iii-iv, *Annual Report 2000-01*: iii). John MacAskill's recollections of his experience of the first community purchase of the 1990s on the North Lochinver Estate is interpreted in the same light, presenting the reversal of a mind-set: the eviction of 'defeatism'. He describes the move from being a tenant to owner and the act of symbolic re-possession as liberating, resulting in 'a spiritual change in the community ... a sudden sense of freedom' (MacAskill 1999: 160). The sense of justice also seeps through his recollections: 'Achieving ownership of the land was the realisation of a dream that the wrongs of centuries past should be rectified' (MacAskill 1999: 21).

Historiographical debates

The discursive issues raised earlier as regards contemporary sources suggest that the clearances offer the historian a feast of diverse material full of contradictory frames of reference, conflicting preconceptions and assumptions, and shifting justifications. Changing conditions and the particularities of individual situations – given the time frame and range of estates involved – also mean wide disparities and evolving viewpoints. If the body of evidence is substantial, it is also plagued by deep polarisation marked by an increasing awareness that past events – particularly trauma – could be used as instrumental tools. As demonstrated, the historical record – events and witnesses – has, over time and over use, also become deeply etched in collective memory, now invoked alongside calls for forgiveness or alternatively redress for past injustices.

Clearances historiography reflects the advances and changes that have characterised history as a discipline since the beginning of the 20thC. Gradually it has encompassed new perspectives and methodologies and has moved towards greater inter-disciplinarity and cross-fertilisation within the social sciences.[23] Very broadly, many Clearances works have adopted a socio-economic approach, first initiated during the interwar period and more fully developed from the late fifties when Scottish economic history expanded, and as already alluded to, agricultural, environmental, but also migration issues were investigated. Using a range of demographic and scientific sources, apart from estate and elite records, socio-economic analyses stressed demographic and exogenous environmental factors to account for the huge outflow of people from the region in the second half of the nineteenth century, irrespective of the actions taken by the lairds (Gray 1957, Youngson 1973, Flinn 1977, Bumsted 1982, Richards 1982). Other analyses have

23 For details about the progress of Clearance research, see Richards (1985: 126-76), Gourievidis (1993: 5-118), Richards (2007: 22-23, 222-25).

been infused by theoretical models, Marxist in particular or Gramscian (Withers 1988) in the case of culturally oriented approaches. By the early 1970s, a parallel academic tradition was emerging, challenging what was perceived and constructed as the orthodoxy in Clearances historical interpretation: the reproduction of classical economic models and a mainly Whig form of Highland history reflecting elite views. Whether drawing from the sociological theory of underdevelopment or inspired by E.P. Thompson's approach of 'history-from-below' in *The Making of the English Working Class* (Carter 1971, Hunter 1976), these interpretations chimed in with the *History Workshop* movement and its radical approach to history; the Scottish peasantry, often giving pride of place to the Highlanders, proved a fertile case study in works stemming from this school of thought (Hobsbawm 1980, Carter 1981a, 1981b).

The historiographical debate which surfaced in the 70s was eloquently and concisely enacted in the academic journal *Northern Scotland* through a series of review articles on books regarding the Sutherland Clearances; two of these articles were penned by Eric Richards (1974-1975) and James Hunter (1974-1975), the third one by a philosopher, Stewart R. Sutherland (1974-1975).[24] What transpired through their arguments was, at one end of the spectrum, the importance of economic analyses which, through close examination of estate papers, in particular management strategies, improvement schemes and financial decisions, drew attention to entrepreneurial efforts and the pressure of complex circumstances – social, environmental, economic; at the other end, their major limitation was stressed as they were accused of reproducing the improvers' standpoint to the detriment of human and cultural dimension. From a philosophical stance, at the heart of the 'improvement or development' debate lay notions of ethics and economics reflecting differing value systems. The indicators used to measure the effect of the Clearances varied depending on the community of interest considered. Estate representatives read way of life in terms of economic performance and viability; within this perspective, other indicators such as cultural dislocation and break-up of social networks assumed the mantle of incidental facts. The nub of the argument was clearly the issue of sources and, in particular, sources bringing out the voice of the people.

Among the sources which afforded insight into the mind-set and life of Gaelic society were those belonging to oral tradition, many of which were recorded and deciphered by Eric Cregeen and members of the School of Scottish Studies in Edinburgh (Cregeen 1974a, 1974b, 1993-1998). Oral history was a new territory from which fascinating aspects were emerging with particular regard to the construction of personal remembrance, its internalised codes and stereotypic forms, its weaving of folk tales with personal tales as well as the importance of the context of recollection. Attention was drawn to the effect of time which could imply shifts of focus and emphasis, along with a measure of re-assessment (Thompson 1988). This applied to recently recorded memories and also echoed the point made

24 The books reviewed were Adam (1972) and Richards (1973a).

by the Napier Commission about the evidence it heard in the 1880s. Most of all, it meant that orally transmitted Gaelic songs and poetic production achieved greater visibility and credence within the corpus of documentary sources. As mentioned in the first chapter, the research of oral historians has proved invaluable for memory studies; new light has been cast on eye-witness accounts of Clearances based on the recollection of events, such as Donald MacLeod's *Gloomy Memories*, initially often dismissed because of its partisanship and adversarial writing context, and now valued because of what it tells about 'how people used memory to come to terms with their present' (Withers 2005: 34).

Found alongside academic developments, the work of popular historians published in the 1960s brought back to life the radical accounts of the nineteenth century often, as in the case of Prebble, borrowing methods of emplotment, characterisation and setting from literature, thus engaging with the readers' imagination. Also buttressed by ideological convictions – socialist or neo-colonialist – these historians' aim was to denounce the action of the powerful and present the crofters' viewpoint by focusing on alternative voices. Their purpose was close to James Hunter's objective, yet in the initial preface of his book, he distanced himself from those authors:

> As propaganda, they were brilliantly successful. As history, however, they seem less than adequate to most modern historians simply because they were written by men who were totally committed to one side in the class struggle between crofters and proprietors …
>
> It is freely admitted that the popular view of Highland history is, as has been said on more than one occasion, invariably prejudiced, frequently emotional and sometimes inaccurate (Hunter 1976: 4-5).

In 2000, in an updated preface, Hunter admitted this position owed much to his attempt at conforming with academic conventions and realigned himself with his predecessors, whilst also asserting the political underpinnings of his work:

> *The Making of the Crofting Community* is openly, and intentionally, *about* the crofting population's long, and still to be concluded, struggle for equity, justice and, indeed, land reform (2000: 17).

In many ways the historiographical debates which are recurrently revived and publicised, are cast following a set polarised model: the contours of political fissure split around the conditions of crofting and landownership.[25] They also resort to universally acknowledged grammars and icons of victimisation and trauma through which the clearances are filtered and recast, such as the Jewish Holocaust or the

25 One recent example is the deliberately inflammatory arguments developed by M. Fry (2005).

shipping of slaves to America (Richards 1985: 167-70, Gouriévidis 1993: 109-10, Asherson 2002: 213-15, Basu 2007: 192-200).[26] From the perspective of post-memory affecting those growing up without individual experience of the events and, by now, with few unmediated tales from descendants, but surrounded by broader cultural narratives of suffering, expatriation, or the damage of ethnocentrism, these offer persuasive and stirring explanatory grids. Basu (2007: 192-200) has shown, for instance, how compelling these templates are for descendants of emigrants in quest of their own family story and a collective 'ethnic' identity.

Both the political polarisation of the interpretation of Highland history and the use of universal vocabularies of victimisation tend to overshadow the continuing progress of research made by social, military, or local history along with historical geography, or cultural and, most importantly, Gaelic studies, which overall help produce a much more complex picture. In it the people's experience has increasingly been given its rightful place, and its many shades and facets have been brought to light. To a significant degree, the works of popular historians and people's history drew attention to controversial or little investigated issues which continue to be explored; among them may be mentioned the alleged passivity of the Highlanders (Richards 1985: 287-369, Withers 1988: 327-401), the pre-clearance period as little altered by new commercial values (Macinnes 1996, Dodgshon 1998, Devine 1999a), the role of local landlords during the great potato famine (Devine 1988), the positioning of the Church – Established and also Free Church after the Disruption – as regards evictions, poverty and protest (Meek 1998, Ansdell 1998, MacColl 2006, Paton 2006), and the causes and motives of emigration from the region over time (Bumsted 1982, Devine 1992, Harper 2003). One interesting development – which has recently attracted media attention – is the reclaiming of Lowland agricultural re-structuring and transformation under the label 'Clearances' (Cowan 1977-1978: 166, 1992, Devine 1994b: 111-64, Devine 1999b: 147-51, Aitchison and Cassell 2003, Devine 2006), which certainly suggests that 'agrarian transformation is not merely Highland and Lowland, but a generic phenomenon in Britain' (Richards 2007: 23), in other words an integral part of the agricultural 'revolution', but also shows the expansion of a symbolically loaded image, once geographically confined to the Highlands and Islands and now applied to the whole of Scotland – albeit with a significant number of grey areas and much qualification (Devine 1994b).

* * * *

Collective imaginaries of the Clearances had long been nourished through oral tradition – given recognition through the official platform of the Napier Commission – literary narratives and political discourses before historians addressed Clearances issues in any depth. The Clearances had already become embedded in Scottish

26 See Hunter's reference to 'a crime against humanity' in the preface to the new edition of *The Making of the Crofting Community* (2000: 6).

culture, stored in the archives of collective memory as a national wound – and a contested one at that – one recurrently used in leftist and nationalist discourse with particular regard to land issues. The Scottish Parliament's debate of a motion for apology in September 2000 may, arguably, be as much a reflection of this long-lasting sense of injury, as a reflection of changing political conditions at home and abroad. Its wording and ensuing debate are an interesting conflation of 'global post-colonial politics of reconciliation' (Basu 2007) and national politics of land reform.

> [Parliamentary motion] That the Parliament expresses its deepest regret for the occurrence of the Highland Clearances and extends its hand in friendship and welcome to the descendants of the cleared people who reside outwith our shores.
>
> [Lewis Macdonald, Labour member for Aberdeen Central] I do not want an apology from this parliament … I want this parliament to build on the resistance and achievement of the past 150 years to deliver a far-reaching land reform that will secure the future of our crofting communities, to deliver a secure future also for the Gaelic language and culture as part of the heritage of the whole of Scotland, and to deliver social justice and economic opportunity, which are the shared ideals of Uist land leaguers and Aberdeen trade unionists alike (Scottish Parliament 2000).

The significance of historical research in forging those imaginaries seems relative and certainly depends on whether fresh analyses conform with this dominant trope of the clearances. From this perspective, historians enter the fray as either 'sanctifiers' or 'disrupters' of memory. It is also a collective memory which is not without mental nor cultural complications, not least for descendants of cleared expatriate Highlanders who might have to negotiate between identifying with the traumatic construction of emigration as forced dispersal – banishment – and the romantic construction of kinship, within a clan whose figurehead may well have helped hatch policies leading to exile. It might simply lead to paradoxical affiliations (Ward 1996).

Museums in their displays manufacturing their own narratives of the Clearances have to take into consideration the force of collective images, the public they target together with more prosaic aspects such as constraints of space and financial means. Most of all they have to clarify their objectives. The following chapter will consider the museums where interpretations of the Clearances appear from the angle of the agencies that sponsor them and their emergence. Chapters 3 and 4 will focus on Clearance narratives and examine how museum curators deal with such a disputed, politically- and symbolically-charged issue as the Clearances, in particular assessing the extent to which their narratives are reflexive and present the dilemmas of interpretation.

Chapter 3
National and Local Museums: The Structural Environment

Responsive to popular feelings, the foundations of many independent museums lie in a recognition of the social and economic change which began between the Wars and has since become more rapid, and of the dissolution of the inherited order and occupations of local communities. The spirit, comparable to that expressed by Walter Scott in relation to the passing of an earlier order, is often as much proudly sentimental as documentary (Museums and Galleries Commission 1986: 9).

The heritage of the Clearances in the national, let alone diasporic, psyche has had far-reaching ramifications, but the extent to which the museum world has grappled with the period and the way it has done so must be related to an array of parameters starting with the nature, purpose and origin of the venues addressing the issue, and extending to external considerations such as economics. The emergence of establishments interpreting the clearances is undoubtedly determined by historical, geographical and, not insignificantly, financial conditions. At a national level, the avenues followed as regards the interpretation of Scotland's past become all the more crucial, as they governed policies of development, such as the type of museum to be established and their location. But museums are also important institutions within civil society as they can be used to legitimise, alter or contest versions of the past as well as the value and power systems underpinning them. This is even more potent in national institutions which can inscribe events, symbols, figures, practices within the narrative of the nation as a whole.

The clearances in historical discourse have not simply been addressed as a regionally- and culturally-specific feature, but have been inserted within national analyses focusing on socio-economic transformation, for instance as one of the facets of the policy of Improvement. Therefore, in theory, representation of the Clearances might be found in a large variety of museums – with an agricultural, regional or social bent, with a national or local focus, and with a long or relatively short-term perspective. Where can visitors find information about the clearances in Scottish museums? Among the nationals, information can be gleaned at the Museum of Scotland (MoS) and the National Museum of Rural Life (NMoRL). However, both are comparatively recent developments, the former opened in 1998 and the latter in 2001. At a regional level, where could visitors go, looking, for example, for a history of crofting agriculture in the Highlands and Islands as a whole or, better still, for an exhibition with an *histoire totale* perspective which would cover the socio-economic, political and cultural past? Beyond the short-lived exhibition outlining the birth of crofting at the Newtonmore branch of the Highland Folk Museum (HFM) in the late 1990s, no comprehensive exhibition has been attempted; the crofting history panels at Newtonmore were abandoned to concentrate purely on 'living history'. The question is slightly easier to answer if the visitor is seeking information concerning a specific area or locality as a variety of local museums address the issue.

The aim of this chapter is to examine the legislation, organisations and specific policies conditioning museum development, together with the 'social arenas' involved- individuals and groups. The discussion will limit itself to structural and social elements mainly relevant to the museums surveyed, as it is not the purpose of this book to analyse the museum sector per se; chapters four and five will return to some of these issues in relation to exhibitionary practices and museological choices linked with Clearance narratives. The heritage venues surveyed come in many forms as do the nature and focus of their exhibitions (see Table P.1). They range from small establishments with local interest and hardly any or no paid staff to major national institutions such as the MoS. All the museums included in the survey except one, the HFM were established after 1965 and the majority of them before the 1990s (see Table P.1). All of them, except the HFM run by the local council since 1974 (Noble 1977-1978, Grant 2007), the MoS and the NMoRL, are independent – managed by trusts or privately owned. One of their most significant differences stems from their sources of funding, and for convenience, they may be distinguished between three different types: national, local authority and independent museums.

National museums

For a collection to qualify for national status, it has to meet the criteria laid down by the Museums and Galleries Commission:

> These are that the collection should be of national importance; that ownership of the collection be vested in Trustees on behalf of the nation; that the museum should be wholly (or mainly) funded from central public sources; and that the government is able to call on its staff from time to time for such expert advice in its field as it may require (*Cultural Trends in Scotland* 1992: 40).

The Williams Report, published in 1981, is one source which provides invaluable insight into the directions adopted for the development of the National Museums and Galleries in Scotland – with regard for the 'continuing pressures on resources' (Williams 1981: ix) – partly shedding light on the issues raised with respect to exhibitions featuring clearances related elements. Firstly, it confirmed the long-mooted idea that a new venue should be found or erected to house the collections held by the National Museums,[1] much of which were stored away from public access. Its

> ... most fundamental recommendation [was] that the artifacts of Scottish culture should be the concern of a new institution, the Museum of Scotland, at least comparable in space, staffing and resources with the Royal Scottish Museum, and wider in scope than the present National Museum of Antiquities of Scotland which it is intended to replace (ibid. Foreword).

The MoS was the offspring of the 'shotgun marriage of two institutions of fundamentally different personalities and philosophies' (McKean 2000: 42). One was the National Museum of Antiquities of Scotland (NMoAoS), whose collections were started in 1781 by the Society of Antiquaries; the other, the Industrial Museum of Scotland, had been established by the Board of Trade in 1854. Their collecting principles, purposes and origins were far removed from each other, yet they shared one critical particularity, neither ever intended to tell Scotland's story. However, when the MoS was conceived, it was meant to showcase Scotland's culture, whilst presenting its history; this was to be achieved through the use of artefacts rather than extended text and reconstruction (R.Anderson 1989). In effect two objectives were fused, as the MoS was to be 'the prime repository for artefacts reflecting the cultural heritage of Scotland' and 'a magnet for visitors ... who want to learn about Scottish history' (Williams 1981: 13). Therein lie the predicaments faced by the curatorial staff of the new museum: how to unite the representation of cultural production with a historical narrative, whose watersheds or critical issues might not be so easily conveyed through material culture; how to reconcile timelines which might not necessarily converge. In the words of Anderson, former director of the National Museums of Scotland, 'the resulting experience will not be balanced history, but that part of the history of Scotland and the Scottish people that can be illuminated by its material culture and natural objects' (R. Anderson 1989). The following chapters consider the ways in which these guiding principles affected narratives of the Clearances.

The national collections contained an extensive selection of country life artefacts, amassed by the NMoAoS since the nineteenth century, and the Williams report also commented on the representation of Scottish rural life. Although it acknowledged the recurrent need for a national folk museum expressed in the evidence it received, it did not recommend:

> that a centralised folk museum for the whole of Scotland be set up ... which would involve the indiscriminate re-erection of buildings or removal of other artefacts from different parts of Scotland (Williams 1981: 16).

1 The National Museums of Scotland was established in 1985 under the National Heritage (Scotland) Act and combined the NMoAoS and the Royal Scottish Museum. See Taylor (2004) for details concerning its background and collecting policies.

Instead, it advocated that a Lowland Museum of Rural Life be established, revolving on a working lowland estate and that the HFM 'be regarded as the main museum centre for Highland folk life' (ibid. 78), providing help and support to more remote venues in the region. The evidence collected and the final proposals shadow previous discussions relative to the best means of representing and interpreting Scottish rural life: through a museum dedicated to Scottish agricultural history; through an open-air museum which would bring together all the small collections scattered throughout Scotland and help visitors form a comparative view of the various faces of Scotland – a open-air museum on the models long-established in Scandinavian countries; or through a network of regional and local country life museums which would delve into more intricate details.

In 1929, the idea of an open-air museum of Scottish rural life was suggested by the Royal Commission on National Museums and Galleries (1929: 60). Over two decades later, reports devoted to the Scottish situation made similar recommendations. In 1951, the Advisory Council on Education in Scotland reporting to the Secretary of State urged, in respect to museum development, that a National Folk-Lore Commission and a National Folk Museum be established. In 1952, the Philip Report on the NMoAoS, deploring the 'cramped' conditions of its displays which thwarted further expansion, strongly endorsed the creation of a public Folk Museum – its shape and remit to be decided (1952: 9, 18-19). But it was only in 1959 that action was taken by the Trustees of the NMoAoS with the creation of the Scottish Country Life Section of the Museum and the appointment of a new Assistant Keeper, Alexander Fenton (Fenton 1985: 2-3, Fenton and Sprott 2002-2003: 3). Although both collections and documentation expanded, exhibition of material was still hampered by lack of space compounded by the fact that the Museum's Trustees had neither the power under their Act to appeal for fund for a building project, nor were they entitled to hold land, a pre-requisite for any open air venture.[2] Yet, the work of the Country Life Section received a boost when, in 1965, the Directors of the Royal Highland Show invited it to set up annual exhibitions on its showground at Ingliston. Finally, in an attempt to circumvent the legal restrictions of NMoAoS' Trustees, the Scottish Country Life Museums Trust was formed in December 1970 as a limited company. The Trust's object was to explore and promote means of exhibiting the NMoAoS' collection of rural artefacts as well as encourage the development of similar projects; a conference, jointly organised with the Countryside Commission for Scotland, was devoted to these topics in 1972. The Trust was crucial in raising money for the establishment of the Scottish Agricultural Museum at Ingliston which, after some setbacks, was finally open all year round in 1980 (Fenton 1973a). When visited in the early 1990s, the museum had a strong ethnological bent and was, above all, concerned with the close interconnection between people and their environment and the evolution of agricultural techniques and practices, using examples from the whole

2 For the background to the creation of the Scottish Agricultural Museum, see Fenton (1978).

of Scotland and featuring regional disparities; it provided very limited insight into socio-economic and political issues affecting the Highlands and Islands and the Clearances were not examined.

The Trust's other concern was to be the development of an open-air museum of Scottish country life and the 1972 conference was used as a brainstorming platform on the subject and the options opened. A decade later, the Williams Report chose to endorse a regional approach to the interpretation of rural life and, in the Highlands and Islands, efforts concentrated on the HFM, the brainchild of I.F. Grant. Pitmain Lodge and grounds at Kingussie had opened in June 1944 and was the third home of I.F. Grant's collection of Highland rural artefacts; on the museum's ground, replica buildings could be visited, such as a blackhouse and water-powered 'clack' or horizontal grain mill from the Island of Lewis, and a cottage from Inverness-shire. The HFM was an entirely private endeavour, I.F. Grant establishing the collection and running the museum by herself until 1954. Because it was privately owned, the museum was not eligible for grants and was not exempted from Entertainment Tax as it was not deemed in accordance with the authorities' definition of 'educational' (Grant 2007: 187, *MJ* 1945). The museum property and collections were then first bought by the Pilgrim Trust before being presented to the universities of St Andrews, Glasgow, Aberdeen and Edinburgh (Cheape 1986: 117). The financial frugality of the 1970s eventually forced the universities to hand over the museum to the care of the Highland Regional Council and, in 1974, 'fulfil[ling] I.F. Grant's dream', it was transferred into public ownership (Noble 1977-1978: 145-46). In the late 1980s, the Kingussie museum was complemented by a larger site, at Newtonmore, devoted to living history and opened to the public in the mid 1990s. It is at Newtonmore that a crofting history display was presented for a few years.

Following the same orientation towards open air display and living history as the HFM, the collection of the museum at Ingliston was eventually incorporated into a new venue, the Museum of Scottish Country Life opened in 2001, renamed National Museum of Rural Life (NMoRL) in 2006. Economic austerity imposed constraints on the form of the new venture; the Scandinavian model of re-sited buildings was deemed prohibitive in terms of creation and maintenance costs and, eventually, the new museum was born out of a partnership between the National Trust for Scotland and the National Museums of Scotland[3]. It combines the interpretation of rural life in Scotland during the period of its transformation – often labelled 'the Agricultural Revolution' – with a working farm bequeathed to the National Trust by James and Margaret Reid and meant to illustrate the farming practices of the 1950s. The displays in the museum building are articulated around three main parts: land, people and tools, and the Clearances are featured under the title 'An emptying land' in the people section.

3 For background information concerning the making of the NMoRL, see Fenton and Sprott (2002-2003).

In consequence, incorporating the Clearances within national displays is a comparatively recent development, however national museums now do sanction their integration within narratives of the nation. If this was not previously the case can hardly be imputed to deliberate omission or avoidance, but to financial limitations, museological orientation – such as the ethnological priorities of Ingliston museum – and lack of an appropriate 'agency' (Ashplant et al. 2004b: 17) at a national level, given in particular the long period of gestation preceding the birth of the MoS or the legal restrictions hindering the actions of the NMoAS' Trustees. The way these national museums incorporate the Clearances in their narratives of Scotland's past and with what effects will be explored in the next two chapters.

Local authority and independent museums

As already noted, the council-run HFM excepted, all the other museums surveyed were opened after 1965. This certainly reflects the pattern of museum provision in the Highlands, but also, more widely, mirrors a nation-wide and UK trend which saw the rise of independent museums since the 1960s (Kavanagh 1990: 47).

Table 3.1 Number of museums in Scotland

1928	1948	1968	1978	1981	1990
67	75	100	130	300	366

Source: Museums and Galleries Commission (1986: 8) and Scottish Museums Council (1990).

In 1977, it was estimated that museums and interpretive centres in the Highlands numbered around 25 (Highland Regional Council 1977: 49); by 1990 the Highlands and Islands as a whole – including the northern Isles – could boast some 70 such venues (Scottish Museums Council 1990). Local authorities have been increasingly instrumental in the expansion of local museums and, over time, became placed under growing pressure to promote and develop the museum sector, as the societal role of museums was redefined to encompass socio-economic vocabularies stretching beyond their conservation and educational functions. The legislative framework which regulated the delivery of cultural services at a local level and was meant to assist museum development through capital or revenue funding, as a result, needs to be clarified.

The initial piece of legislation giving local authorities power to establish museums and galleries dates from 1887, the Public Libraries Consolidation (Scotland) Act, but museum development was hindered by lack of financial provision as they were often seen as mere adjuncts to libraries. The Act was subsequently revised, most notably following the progress of local government reform. In 1973, this power was transferred to Regions, Islands and Districts (Local Government (Scotland) Act) and, in 1982, it was restricted to Districts and Islands alone (Local Government and Planning (Scotland) Act). Section 14 of the Act placed a duty upon these authorities 'to ensure that there is adequate provision of facilities for the inhabitants of their area for recreational, sporting, cultural, and social activities'. The power given to local authorities was discretionary and, by 1986, the record of local government contribution to museum development was deemed patchy and uneven, the report of the Museum and Galleries Commission (Miles Report) (1986: 17-18[4]) underlining that 'examples of a District Council or a Regional Council putting funds into an independent museum trust are very rare and their initiative in doing so is exemplary' and blamed the imprecision of the wording 'adequate provision'. It recommended that local authorities make curatorial staff available for help and advice and be prepared to assist museums financially. It also singled out Caithness for its pro-active and innovative actions (ibid. 28). Although the plans by the Highland Regional Council to set up a team of curators providing advice and co-ordinating the activities of existing museums in the area were cut short by the 1982 Act, a year later Caithness District Council appointed a museum professional offering curatorial assistance to independent museums and, in 1985, established the first District Museums Service in the north of Scotland. It was shortly imitated by Ross and Cromarty and, in 1987, Skye and Lochalsh (McCann 1986: 19). Unlike more compact urban authorities, it made little sense for district councils in the Highland area with their scattered and isolated communities to run museums; instead they contributed to their development and survival through guidance as well as capital and revenue grants. *Dualchas*, Skye and Lochalsh Area Museums Service, was described as a 'museum without walls' helping 'communities, local history societies and existing independent museums' with the interpretation and preservation of heritage (Carmichael 2002: 113). Ross and Cromarty Council, which facilitated the refurbishment and radical transformation of Ullapool Museum and Visitor Centre (Ullapool M&VC) and subsequently aided with running cost, is another case in point as it never managed a museum, acting as an 'enabler rather than provider' (G.Watson 1990: 17) – a position which, willy-nilly, was well suited to the dominant ethos of the 1980s and 90s. The Islands, run as single tier authorities, each had their own directly-managed museum service, also providing advice to the independent sector.

By the late 1980s, improvements were reported throughout Scotland with 70 per cent of district and islands councils running professionally staffed museum services and another 9 per cent offering significant support (Ambrose 1989b: 1). In the Highlands and Islands, in 1993, a year before the Local Government etc. (Scotland) Act established new unitary councils, a report on museum development listed the districts devoid of museum services and, in some cases, budget – Badenoch and Strathspey, Lochaber, Nairn and Sutherland – and summed up the main challenge

4 See also Ambrose (1996).

facing local actors: 'fragmentation on a grand scale' (Broad Oak Consultancy 1993: vi, 19). In the short term, the situation was remedied with the appointment of a museum development officer, initially on a temporary part-time contract, a post jointly funded by HIE and SMC (*MJ* 1995: 8, SMC et al. 1999: 5). Even though advice and curatorial support were more accessible, there still were substantial funding disparities between private sector museums with some receiving hardly any or no funding at all. The pages of the *Press and Journal* provide an insight into the precarious and volatile funding situation of independent museums in the northern Highlands in the late nineties. According to the reporter, not only were museum representatives alarmed at prospects of cash shortages, but they complained of lack of consultation on the part of the council about to divulge plans regarding their future (*PJ* 1997). The situation was later compounded by reports of unfair treatment, Timespan and Strathnaver Museum said to be receiving 'a mere pittance compared to [museums] in Ross and Cromarty' and there were calls for equalisation grants (*PJ* 1998). Outside the Highland region, Arran museum was not faring much better, as the annual grant it used to receive from Cunningham District Council ceased at the time of local government reform. Confusion and uncertainty seemed to dominate the museum profession in Scotland at a time which saw the effect of local government reform dovetail with the changes ensuing from devolution, as responsibility for culture was devolved from Westminster and the Scottish Executive was deciding on a national cultural strategy. Simultaneously, the Labour government introduced the concept of Best Value, placing on local authorities a duty to deliver services to clear standards – in terms of cost and quality – by the most economic, efficient and effective means available (Vize 1999); it was encapsulated in the '4 Cs' – Challenge, Compare, Consult, Compete (SMC 1998: 3). At the dawn of the new millennium, Best Value was a central tenet of the recommendations made in a SMC report assessing museum provision in the Highlands; discrepancy in the funding of independent museums, over-reliance on volunteers – a feature repeated throughout Scotland (SMC 2002: 5) – vulnerability to economic fluctuations, but also good level of service led to the conclusion that a 'baseline level of funding …' was required. It was to be '… implemented through the use of service agreements …', related to performance and standards, '… an approach based on the principle of reciprocity' (SMC 2001: 1-5). Best Value was also central to the Local Government in Scotland Act 2003 which placed on local authorities a statutory duty of community planning and well-being; in term of cultural services, this meant developing community-oriented policies, including all museums within their area. Since then, in the Highland Region, registered museums entering a service agreement have been able to receive revenue funding; it is the case of Strathnaver Museum, Mallaig HC, Ullapool M&VC and Timespan.

Looking back over the period when the museums surveyed were established, it seems that, since the changes which altered their powers in the mid-seventies, local authorities have assumed enabling functions rather than pro-active roles in the creation and survival of many venues. They have helped kick-start long-standing projects, through the provision of one-off grants complementing donations and fund-raising activities. It is a recurrent feature in the history of most of the venues surveyed, as the following examples demonstrate. In the late seventies, Arran museum received several grants from Cunningham District Council to purchase some of the buildings now part of the museum complex; Strathnaver Museum received assistance for the renovation of the Church it was to occupy from Caithness District Council, as did Clan Gunn Heritage Centre. Timespan obtained a substantial contribution from the Highland Council in the mid-eighties. Over time in view of the rapid growth of museum projects – Scotland was said to have 'nearly twice as many per head of population as most other parts of the United Kingdom' (Museums and Galleries Commission 1986: 125) – the priority became one of improvement of standards rather than new developments, and local authorities were encouraged to adopt more careful regional planning in order to avoid duplication or unsustainable projects. In some measure, the Highlands and Islands Museum Forum, created in 1989 as a means of facilitating cooperation and exchange between museum members, was meant to offset these problems, promote cohesiveness and give museums a voice. In addition, following the Thatcherite ideology which dismissed local authorities as the most efficient means of delivering services and the economic stringency of the eighties and nineties, they had to scrutinise their expenditure on private sector museums – and more generally the museum sector as a whole – and promote plural funding or partnerships, enticing independent museums to derive some of their income from business sponsorship. With Best Value, New Labour did not significantly alter the course of economic circumspection and public accountability, two of the principles embodied in John Major's Citizen's Charter and directing local authorities' involvement in the cultural sector.

Sources of additional funding

Besides local government, heritage projects could also hope to muster support from many other sources, in particular the growing number of quangos which, since the 1960s, have been a trademark of British governance whose prevailing stance was the delegation of responsibilities to non-elected government-appointed bodies. Each have a clearly demarcated remit and those that have made an impact on heritage development in the Highlands and Islands may generally be regarded as having cultural, environmental or economic concerns, some specifically targeting regional development. In the main, they have some advisory and mostly enabling functions – through the provision of grant aid – and their specific sphere of activity directs their vision of the role of heritage and conditions their financial help.

The organisation most closely related to museum development is, of course, the Scottish Museums Council (SMC) which was founded in February 1964 as the Council for Museums and Galleries in Scotland – one of the Area Museum

Councils in the United Kingdom.[5] In form, it is not quite a quango since its board is elected by the membership and often includes local government members. Its brief was to raise the general standard of local museums and it acted as the main channel for central and, after 1998, devolved government funding to non-national museums, providing a variety of support services, inclusive of grant-aid, for its membership – independent museum trusts, local authorities, development corporations, universities, regimental museums, historic houses and other allied organisations (SMC 1988-1989: 2); it has also acted as adviser and lobbyist for the museum sector. Below are figures provided by the SMC listing their local authority and independent sector members over time. However, it is essential to bear in mind that discrepancies with Table 3.1 can partly be attributed to the fact that not all independent museums were and are SMC members; additionally, inconsistent figures as regards local authority museums in the 1990s are due to local government reorganisation and its subsequent impact on the museum world.

Table 3.2 Breakdown of SMC membership – Local authority (LA) and Independent (Indpt) museums

1968		1978		1988		1997		2008	
LA	Indpt	LA	Indpt	LA	Indpt	LA	Indpt	LA	Indpt
30	11	50	41	51	88	32	137	32	159

Source: Data from Annual Reports and Director's Reports. I am grateful to Mark Macleod from Museums Galleries Scotland (formerly SMC) for compiling these figures for me.

In keeping with UK policy and the recommendations laid down in the Miles Report, alarmed at the uncontrolled and anarchic growth of independent museums, the SMC has operated a voluntary registration scheme, introduced in Scotland in 1990, which guarantees appropriate professional standards and has become a major eligibility criterion used by local councils and other funding bodies for awarding support (R. Clark 1993: 18, Ryder[6] 1995: 2-3). The SMC also co-ordinates between grant-providing bodies, whilst offering advice to bidders; this dual function transpires in its relation to the Heritage Lottery fund: acting as adviser to new applicants and helping it assess the bids (Ryder 1996: 8-9).

On the environmental side, the Countryside Commission for Scotland (1968-1992) – whose functions were subsequently taken over by Scottish Natural

5 It changed its name to Scottish Museums Council in 1984 and in February 2008 was renamed Museums Galleries Scotland. See Ambrose (1996), director of the SMC from 1986 to 1994, for details of its development and function as well as an assessment of its role in the expansion of museum provision in Scotland.

6 Jane Ryder was Director of the SMC.

Heritage (SNH) – fostered interpretive approaches devised to make the visitors aware of all aspects of the local environment surrounding them, of the character of specific areas. The Commission was particularly sensitive to the nature of interpretation and its design, advising and running training courses on interpretive planning and other related issues. Under the 1967 Countryside (Scotland) Act, it was able to provide grant-aid to recreational and educational projects and singled out heritage venues with a folk life focus. As one of the main actors behind the discussions regarding country life museums in Scotland in the early 1970s, it was already sounding alarm bells as regards the danger of duplication and the need to adopt sound regional interpretive strategies and even consider alternatives to the establishment of open-air museums (Aldridge 1973, Fraser 1982-1983: 3, Museums and Galleries Commission 1986: 108).

Heritage is also promoted by a variety of agencies broadly linked with the economic sphere. Concentrating on the improvement of the social and economic conditions of the Highland and Island region, the Highlands and Islands Development Board (HIDB), created in 1965 – Highlands and Islands Enterprise (HIE) from 1991 – was meant to tackle the twin problems of depopulation and lack of economic diversification which had blighted the development of the region. It has approached heritage from an economic perspective linked with tourism, which was one of the touchstones of its initial policies, and as a means of promoting social regeneration through community-led initiatives. The vast majority of venues surveyed has benefited from the assistance of this agency, whether it be through the HIDB or the Local Enterprise Company (LEC) network part of HIE – some like *Aros* receiving substantial help.[7] Funding is also available from another enterprise quango, the Scottish Tourist Board set up in 1969. Under Section 4 of the 1969 Development of Tourism Act, the STB was entitled to financially support heritage projects regarded as tourist attractions enhancing the visitor's experience – a privilege which its English counterpart lost in 1988.[8] For museums established in the late seventies and the eighties, another source of assistance was the Manpower Service Commission (MSC) which was disbanded in 1988, and on which independent museums in particular had come to depend for staff employed on a variety of short-term projects (*Cultural Trends* 1989: 3). According to the survey undertaken by the Miles Report:

> In the middle of 1985 approximately 40% of the paid staff working in non-national museums and galleries in Scotland, were employed under various MSC schemes. The gross cost of these MSC funded posts must be in excess of £3m in a full year (Museums and Galleries Commission 1986: 113-14).

---

7 HIDB, *Annual Report* (1992-1993: 26, 40), survey answers. Incidentally, in 1975 the area covered by the Board was extended to include Arran and Bute because they shared similar problems to other islands within its territory.

8 Museums and Galleries Commission (1986: 12-15, 108-112), *Cultural Trends* (1989: 1, 3), Fladmark (1994: 13-15, 17-19).

More recently, heritage actors have been able to turn to further sources both national and supra-national. From 1994, the Heritage Lottery Fund (HLF), set up by Westminster to assist a wide range of projects linked with heritage in its broadest sense – buildings, objects, environment and projects enhancing people's quality of life – became a major player in the development of heritage projects (Ryder 1996: 8-10) in spite of initial allegations of elitism and misgivings about the transparency of its objectives and procedure along with the accountability of its modus operandi (*Cultural Trends* 1995: 1-27). It contributed some 50 per cent of the £14.5m needed to fit out the exhibitions in the Museum of Scotland project, leaving the museum to raise the reminder, the basis of HLF funding revolving round partnership – not necessarily matching funding. Much smaller venues in the survey, like *Seallam!* Visitor Centre successfully applied for Lottery funding. Another contributor is the European Union whose ERDF structural funds and LEADER programmes, since the nineties, have made subsidies available to a small number of ventures; funds granted to museums related to the promotion of rural tourism and to the development of isolated communities (Ambrose 1992a). Glengarry Heritage Centre, in 1998, was a perfect match for the LEADER II programme; it was partly funded by HIE network with match-funding from Europe (*Lochaber News* 1999: 8). However, since the turn of the twenty-first century and the widening of Europe's borders, the Highlands and Islands have no longer been part of the priority regions for ERDF – designated Objective 1 area – and financial support has been re-defined (*MJ* 1998: 7). Finally, a prominent feature of arts funding, started in the eighties, was the encouragement and growth of business sponsorship and donations that various governmental measures and policies were designed to foster (*Cultural Trends* 1991: 34-40). Although not a salient aspect of the museums surveyed – nor of museums in general as Timothy Ambrose deplored – it was advocated by the SMC which offered assistance on the subject (Ambrose 1990a). Interestingly, when created, Timespan – whose display used to have a section devoted to the oil industry – was awarded £5000 by Britoil, and a further £1000 through the Beatrice Oil Fund Trust (*Northern Times* 1987b: 9).

Several points might be outlined as regards the structural background of museum emergence and provision. For the great majority of museums surveyed, national or local, plurality of funding has been a fact of life. For its creation, the Museum of Scottish Country Life (later NMoRL) appealed for finance from the HLF, ERDF, the Lanarkshire Development Agency and Historic Scotland, to name but a few of its sponsors. Additionally, given the primacy of market ideology since the 1980s, independent museums even more so than other types of museums have had to rely on plural sources of funding to exist and survive; they have had to adapt to an increasingly competitive environment, in particular by adopting a business-like attitude, factoring in income generation, visitors' satisfaction and commercial sustainability, in districts where they often contend against similar venues for support and rely – in some cases almost exclusively – on admission fees and therefore tourism for existence. In the process of bidding for funding to multiple agencies, they have had to meet differing and, at times, competing criteria and to develop opportunistic vocabularies and projects. The naming of museums itself is testimony to these pressures, the terms 'museum', 'heritage centre' or even 'heritage museum' not necessarily conveying meaningful methodological differences, but reflecting an attempt at conforming with the expectations of a potential source of funding or at deploying a label attractive to visitors. Whilst the necessity to satisfy the needs and requirements of various stakeholders may very well have impacted on the design and content of the exhibitions surveyed, it would be hasty to conclude that this is a main source of influence. In particular, to a very great extent, the museums surveyed rely on armies of dedicated and enthusiastic volunteers to exist and have often developed out of a community of interest which might well reach beyond local or regional perimeters to embrace wider historic connections as with clan museums. This community of interest is reflected in the narratives they produce.

Independent museums and a tradition of self-help and volunteering

Important though it seems, the role of local authorities appears to have had limited impact on the actual conception of the museums surveyed; the main impetus which drove their foundation was not institutional, but came from grass-root enthusiasm and commitment, as recurrently stressed in survey answers, a point more generally acknowledged by Highland Region in 1977:

> The Council recognises the importance of local museums and welcomes the responsibility undertaken by voluntary management committees in the establishment and management of such museums (Highland Regional Council 1977: 49).

The motivation behind museum creation, as the answers to the second survey question indicate (see Appendix), is geographically and historically specific. For many early venues, it was felt that there was a lack of information locally, the respondent for Strathnaver Museum adding that, when they first opened, 'the story of the clearances was being told nowhere'. Many alluded to the increasing importance of heritage interpretation and frequently linked their projects to demand, more specifically demand arising from the interest of visitors after the organisation of temporary exhibitions; from descendants of emigrants in search of their roots; from previous residents who had moved away and; of course, in the case of clan museums, from members dispersed world-wide, thus providing them with a base for research, meetings and activities. Very few mentioned the economic potential of such an enterprise, but one which did, Timespan, hinted at some of the issues that were facing many Highland and Island communities in the 70s and 80s: 'we wanted to slow depopulation by attracting visitors to the village'. This was an objective shared by the Clan Donald Lands Trust as regards the Sleat peninsula on Skye. The Trust was set up in 1971, when the last remaining lands to be held by one of the clan chiefs came up for sale. One of its founding aims was the creation of a Clan museum (now Museum of the Isles), part of a plan to revitalise the local economy, in particular through tourism (Survey answer, Weir 1972). The genesis of some of the venues surveyed brings to light the concerns which propelled their creators. They are primarily related to collecting material traces of the past or salvaging lost memories or untold histories from the trappings of oblivion, particularly if the story speaks to the present; this is frequently combined with reviving a locality's economy through the potential offered by cultural tourism and, in particular root tourism; finally the development of a cultural platform may also be seen as a conduit reinforcing a community's sense of belonging through interaction and shared activities.

Many of the earlier museums surveyed were born of the sheer dedication and energy of a single individual with an interest in regional ethnology and vernacular architecture. Before the second World War, the trailblazer of this movement is I.F. Grant, whose tenacity commands admiration when she recalls that help was far from forthcoming and circumstances, such as lack of good transport infrastructure, turned the very act of collecting into a form of crusade. She watched with dismay the loss of the material heritage she had wished to rescue, just as assistance was becoming available and attitudes were changing.

> I was hampered at every turn with lack of finance. I tried very hard to obtain a grant of some kind but nothing was available. When I think of what could have been done in those days with the facilities now available when it is almost too late I feel disgusted (Grant 2007: 65).

In the 60s, the fledgling folk museums, which were to imitate the HFM on Skye, also had to rely on personal initiative and the generosity of local donors and crafstmen. Skye Cottage Museum, as Jonathan MacDonald's museum was first known, was opened during Skye week in May 1965, an opening performed by Seton Gordon, the Scottish naturalist, who lived very close to the new museum. Jonathan MacDonald had been collecting Highland relics for several years and they were exhibited in a derelict cottage which the owner had acquired, re-thatched and restored with a view to depicting living conditions and building techniques which were fast disappearing as the Hebrides were modernising (*Press and Journal* 1965). Similarly, Colbost Folk Museum also privately owned, was one of several cottages restored by Peter MacAskill, his endeavour described in the *Press and Journal* (1978) as 'an attempt to rebuild the past'.

Some venues stemmed from the vision of one individual surrounded and assisted by the local community. The idea of a museum on Arran was floated as early as the 1920s but never materialised, it was rekindled in 1969 by Bess MacMillan and, eventually, in 1973 when she served as Chairperson of the Isle of Arran Tourist Association, the Arran Museum Committee – later replaced by the Isle of Arran Museum Trust – was formed. It included the archaeologist Horace Fairhurst, who subsequently became president of the museum trust, and was already well-known in academic circles for his excavation and interpretation of pre-clearances settlements in Sutherland; he was later to write a book on the archaeology of the island. In order to raise funds for the project and publicise it, volunteers led by the resolute and inspiring Bess MacMillan organised many social events as well as several open days, during which visitors could appreciate the progress of the endeavour and watch demonstrations of shoeing in the smiddy. The museum was formally opened in July 1979, 'fulfil[ling] a ten-year dream' (*Ardrossan and Saltcoats Herald* 1979), an inauguration attended by representatives of many of the agencies which had contributed to its funding (ibid. 1976, 1977, Small 2002). Strathnaver Museum was also a long and protracted project which started when the local church building became redundant in 1957. Ian Grimble, who lived locally, suggested it be turned into a museum because of its proximity to the ancient Farr Stone and also, conceivably, because, from its pulpit, the controversial Rev. Mackenzie preached his sermons during the Sutherland clearances. A body of local trustees was formed who, in 1960, approached the Kirk Session with their plan to create a museum,

> in which will be preserved the evidence of the history, way of life, and arts of our forbears …
>
> We believe generations to come will deplore our folly if we do not use the old parish church for this purpose. But it will not be possible for them to reproach us also for failing to hand on what remained of our heritage to them (Strathnaver Museum Trustees).

Once handed over to them, it took the trustees many years to amass the necessary funds – and surmount a number of hiccups – for the restoration and adaptation of the building. Horace Fairhurst's excavation of nearby Rosal and other pre-clearance sites in the late sixties, gave the future museum an added boost as it was able to gather the small items found on sites, transforming it into the centre for the interpretation of some of most notorious episodes in the Sutherland Clearances. Caithness District Council is also credited for re-energising the project after local government reorganisation as it helped with building repairs and, in fact, the opening ceremony was performed by the Chairman of Caithness District Council in July 1976 (D. MacLeod 1976, *Northern Times* 1976b, *Scottish Farmer* 1978).

Others are community projects driven by a mixture of economic and cultural motives. The story behind Timespan's creation is recounted in the *Northern Times* (1987b: 8). Intimately linked with the regeneration of the village, five years before the opening of the museum, a committee – Helmsdale and District Action Committee – was formed to 'do something about the village'. It approached the Highland Regional Council that had recently purchased a property, which the committee hoped to convert into a museum. It was put to them that 'they think big' by the Director of development. The centre was an ambitious venture as it was to cost about £200,000 overall. By contrast, Ullapool M&VC grew out of local commemorative activities, namely a temporary exhibition, held in the former Parliamentary Church, for the village bicentenary in 1988. Its success prompted the local volunteers involved to transform the building into a permanent museum for the Lochbroom area. Interviewed by the press, the Trust's chairman described the venture as 'a long, hard struggle … but we've got it by the perseverance of board members, the enthusiastic support of all the trust's members, the whole community and, most important, the faith put in us by all the funding bodies' (Morrison 1995). Like Timespan, a few years before, Ullapool M&VC received financial assistance from numerous sources.

Much of the drive behind the burgeoning of independent museums originated from individuals or groups outside the realm of the state, freely associating to advance common cultural interests, in particular the dissemination of a shared heritage – typical of 'new sociations' (Urry 1996: 59) – and developing a network of 'friends'. A few are also the products of single individuals devoted to the preservation of local history, artefacts and vernacular architecture. At times single-minded and passionate, those venues might reveal highly personalised agendas, tracing, through the narratives of the past they choose to present, the contours of debates and controversies inherent to local communities. The Clearances associated with the history of crofting and its development, because of their political charge, are fertile ground for interpretations frequently promoting values and positions laying bare present concerns and sponsoring reformist and radical wisdom. Recalling the view expressed in the epigraph to this chapter in the Miles Report, weighing the 'spirit' which might have been the driving force behind their interpretive choices of the Clearances made by the social groups involved in the museums surveyed seems absolutely essential; the following chapters will endeavour to do so, starting with an analysis of Clearance narratives.

Chapter 4
Museum Narratives of the Clearances: Themes and Approaches

Personalities, events and ideas are the stuff of traditional history, but what of the reality of Bannockburn, the complexity of David Hume's thought and the tragedy of the clearances? They may be only inadequately conveyed by objects associated with the individuals and historic events concerned. On the other hand, quite simple objects can inspire an imaginative understanding of the realities of everyday life, even in the distant past, which is not readily communicated by the printed page (Mark Jones 1995: 26).[1]

Museum curators, in devising interpretations of the Clearances, have to take stock of a wide range of factors, not least their legacy in collective imaginaries and the expectations these might well raise in the visiting public, from within or beyond Scottish shores. Furthermore, museum narratives, wherever found, often reflect modern concerns whether these are linked to local, national or global issues. Consequently, the way narratives of the Clearances are emplotted in exhibitions depends greatly on the focus, size, scope – transnational (clan museums), national, regional, local – and objective of each venue; not least among these is the time-scale covered within an exhibition and more importantly the way in which they address social time with long term (*longue durée*) or more short term approaches (see Table P.1: focus). This and the following chapters will outline some of the issues raised by museum representations of the Clearances.

The choice of interpreting the Clearances warrants examination and two of the questions included in the survey concerned the importance accorded the period of the Clearances in the displays and the reasons for its inclusion (see Appendix). Out of nine answers, one, Strathnaver Museum described it as 'extremely' important, four, Clan Gunn Heritage Centre (Gunn HC), Timespan, Ullapool Museum &Visitor Centre (Ullapool M&VC) and Glengarry Heritage Centre (Glengarry HC), as 'very' important, Glengarry HC adding that it was 'one of the key elements in the district's pre-20thC history', and Mallaig Heritage Centre (Mallaig HC), as 'an important secondary theme'. The reasons for covering the topic are frequently predicated upon the vision of the period as a defining moment affecting the localities concerned, mainly in social terms, with particular regard to population migration:

---

1 Mark Jones was Director of the National Museums of Scotland and was presenting the MoS project.

> *Strathnaver Museum*: Has affected way of life drastically. Many visitors from overseas come to see the land of their forebears who were evicted during the clearances.
>
> *Gunn HC*: Many of the clan were cleared from Sutherland and Caithness.
>
> *Timespan*: Most important topic in local social history.
>
> *Glengarry HC*: Links with Glengarry County, Ontario, Canada which was settled by Glengarry emigrants in the 18th century with other Gaelic-speaking Highlanders. Also interest expressed by other emigrants notably Antipodes.

Both Ullapool M&VC and Mallaig HC account for its inclusion in displays with reference to the availability of (written) contemporary documentary evidence:

> *Mallaig HC*: It was a major and well-documented event in the area which we felt able to interpret with available resources.
>
> *Ullapool M&VC*: It happened here, the Napier Commission took evidence here.

For the Isle of Arran Heritage Museum (Arran HM), where the Clearances display was a late addition to the exhibition (see Table P.2), the answers given revolve on audience/customers' demand and reaction, the museum not initially intending to make the Clearances a permanent fixture. Expanding on the importance of the period in the display, the informant underlined visitors' response: 'The clearances exhibit draws a lot of interest'; it is part of the exhibition because 'after a temporary display was so well-received, it was decided to create a permanent display.' The museum's decision and aim in this case were a clear response to public demand.

The respondent's description, for Mallaig HC, of the Clearances as a 'secondary theme' is echoed in slightly different terms in two other venues whose informants chose not to lay particular stress on the importance of the period, the Museum of the Isles (MoI) and *Seallam!* Visitor Centre (*Seallam!* VC). The former's answer is directly linked with its objective and *raison d'être*: to portray the past of a global community – Clan Donald – and explain its present-day dispersed nature: 'The Clearances are integral to the main topics covered'; they were included because they 'follow in sequence with the rest of the display and were required from a chronological viewpoint to put into context the info on 'landlords' and 'emigration' etc. The Clearances seem to feature as a matter of context, much like a linguistic signal – a coordinating marker in the narrative. In the case of *Seallam!* VC, the respondent implied an inescapable factual reality, stating that both its importance and inclusion in the display were due to it 'pertaining to island history'.

Turning to the actual narratives of the Clearances, three main strands, tend to guide readings of the period in the exhibitions surveyed with, inevitably, a considerable amount of overlap dependant on the museum's purpose and circumstance: the local, national or emigratory dimension of the process. Only one, Colbost FM, can be said to fit neatly into one category in view of its small size and highly individual presentation.

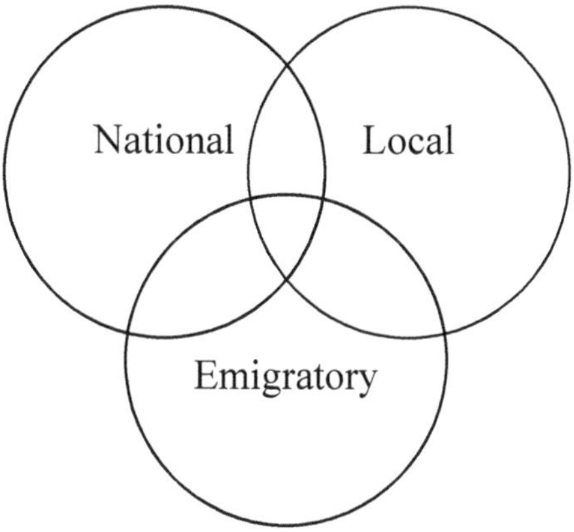

Figure 4.1 Narrative strands

These three strands will structure the organisation of this chapter, museums being grouped according to the strand dominating their narrative of the Clearances. (The main characteristics of Clearances exhibitions are summed up in Table P.2). Not unexpectedly given the fact that the majority of museums surveyed have distinct locational terms of reference, the most prominent strand concerns the local impact of the policy; thus exhibitions might encompass looking into evictions, resettlement policies and their consequences, along with the development of crofting agriculture. The question of relocation is frequently linked to the very existence of the locality where those museums are found and the economic activity and objective from which settlements sprang and expanded: at Helmsdale, Mallaig and Ullapool, museums focus more specifically on the development of their respective harbours and fishing, at Kilmuir (SMoIL), Colbost and Portree (*Aros*) on crofting. In the national strand, museum tend to frame interpretations within the wider register of national policies and trends and might describe the economic, political or intellectual context conditioning the orientation and management of local estates along with the social impact of agrarian transformation. Finally, the emigratory strand also places a high premium on the fate of those who left Scotland as a result of the Clearances, its plot following emigrants to their destinations and incorporating depictions of their journeys and their new lives as settlers. What is of particular interest, in all strands, is the way in which narratives define clearances as a process, and how this is achieved by, for instance, putting specific townships, episodes and individuals under a microscope, or preferring a wide angle view scanning socio-economic organisation and change, put more succinctly by concentrating on the particular or the general. Another important aspect, given the contested nature of the issue, concerns whether museums build reflexive interpretations – interpretations in which a sense of the dilemmas faced by curators is conveyed.

Narrative strand i: Local

The majority of museums surveyed can be bracketed together within this strand, though, (refining the classification further) it is possible to also differentiate between those giving pride of place to evictions and their immediate aftermath and those where the experience or trauma of expulsion is offset by the representation of acts of resistance during the Land Wars and their political consequences. Perhaps unsurprisingly given the geography of confrontation, the latter museums are found on Skye. Whereas earlier protest was more evenly spread, the Land Wars were more intense in the north-west Highlands and the Hebrides (Withers 1988: 351-91). Modern literary imagination replicates a similar geography with novels centring on the tragedy of dispossession and on the dislocation of close-knit communities located in Sutherland, the Spey valley or Shetland whilst those focusing on crofters' resistance and more specifically Land War episodes set on Skye.[2]

Eviction and resettlement

The displays of both Strathnaver Museum and Timespan focus on the first phase of the Clearances on the Sutherland estate (1813-1816) with the controversial and widely chronicled Kildonan and Strathnaver evictions. In both venues, the bulk of the informational content is gained from scripted panels. At Strathnaver Museum, which is more specifically devoted to the Clearances period, it is conveyed with colourful graphic lettering – with its subliminal message – and a presentation relying on archaeological data and artefacts found on cleared sites, one of which, Rosal, was initially excavated by Horace Fairhurst in the sixties.

2 See Chapter 2.

Figure 4.2 Strathnaver Museum, Bettyhill, Caithness
Source: Laurence Gouriévidis 2007.

Additionally, visitors are encouraged, on leaving the museum, to visualise for themselves the ruins of townships now interpreted along the Strathnaver Trail, some of which like Rosal and Achanlochy had long been deciphered for the public (see next chapter). In contrast, Timespan complemented story-boards with several tableaux with life-size models, of which one of the most vivid illustrated 'the year of the burnings' (Prebble 1969: 71) and staged an eviction complete with sound-effects, featuring the Sutherland factor and sheep-farmer, Patrick Sellar, terrorising the Chisholm family into abandoning their home and having it set on fire; another represented the Countess of Sutherland planning improvements. Leaving aside, for the time being, the impact of Timespan's dramaturgy on interpretation and visitor's perception, both narratives present numerous parallels in terms of conclusions and emphases.

Both feature evocations of brutal eviction scenes and insist on the tenantry's plight after the clearances, their declining circumstances sharply contrasting with their poor but contented life before improvements. Although both also make references to the substantial financial contributions made by the Sutherland family to the development of local infrastructure (Timespan) or to relief during the potato famine (Strahnaver Museum) and to their expenditure on the transformation of the estate, the keynote in both narratives is the proprietors' overwhelming commercial motives and determination to turn their estate into a profitable enterprise. What underpins their readings of the period is a critique of the implacable implementation of an economic system provoking social ravages, and a cultural assessment highlighting the discrepancy between the norms and values of the landowners and their peasantry and demonstrating a yawning gap in terms of *mentalités*. To maximisation of profit and private ownership of property is opposed a sense of kinship and belief in customary rights of access to land. Strathnaver Museum, when outlining assessments of the policy, splits them into two deeply polarised view, one of which lists:

Life after the evictions – one view – the factors viewpoint:

1. They are transformed
2. They have begun to cultivate their lots with industry
3. Many of them, with great boldness, have taken to catch cod and ling …

The transformation expected, once the ambiguity of the first statement is raised, not strictly related to land and economic production, but encompassing human character and disposition. In both exhibitions, racial overtones transpire through the words of improvers, as apparent in the following quote in Timespan:

> … to the immensely civilised Marquess of Stafford, his wife's Sutherland tenantry must have seemed as barbarous as Amazone tribesmen … Their lives were, in the final analysis of the Marquess, of unmitigated sloth and over-ripe for improvement.

They bring to light the contemporary perception of the Highlander as an internal 'other' likened to the indigenous populations of distant territories that western nations were fast colonising.

Both offer direct appraisals of the period provided by controlling and intrusive curatorial voices, directly commenting on events and context, using quotes as would historians when introducing the voices of historical actors. Yet, At Strathnaver Museum, story-telling is in the shape of a frequently schematic, compact and expository text, the curatorial voice swiftly moving between topics and viewpoints, whereas Timespan's narrative, in true popular history fashion, akin to creative writing and not unlike John Prebble's style, enlarged on individuals' thoughts and feelings, thereby engaging the imagination and emotions of the visitor/reader. Timespan's engagement with imagination also served a wider ideological vision explicitly meant to echo down to the present, a vision which came to light when the policy of improvement was defined:

> Improvement was the word on the lips of everyone with a sense of enterprise and then as now, as will ever be, improvement was intended to be of prime benefit

> to the entrepreneur who first thought it out. Human life was cheap and plentiful and put on this earth to help not hinder enterprise.

Sutherland evictees, whose fate was compared to that of the human fodder which drove industrialisation forward in the Midlands and in Glasgow, epitomised the numerous casualties of the new capitalist ethos. What also distinguished Timespan was that it dealt at some length with the issue of the resettlement of the displaced, presenting the rapid expansion of Helmsdale in the nineteenth century as one of the most successful projects engineered by the Sutherlands. Yet whilst this success was said to have resulted from the relocation of cleared inhabitants and the growth of the herring industry, the casual mention of the Staffords' financial investment underplayed their contribution to the overall planning and vision. Of the transformation of the region, the abiding memory which lingered from both the exhibition and its accompanying video, was one of human trauma – 'the epitath of the people and their way of life ly[ing] in the empty glens of Strathnaver and Kildonan and the tiny village of Badbea'.

Like Timespan, the exhibitions at Ullapool M&VC and Mallaig HC also devote substantial space to exploring the development of the harbours where they are located. However each weave the Clearances and their local impact into their narratives very differently, partly because of the origin of each town; Ullapool was established as a fishing station by the British Fisheries Society in 1788 and Mallaig grew as part of the agricultural reorganisation of Lord Lovat's estate and his improvement schemes in the nineteenth century. Beyond distinct historical bases, disparity is most of all due to differing thematic organisation of displays and narrative strategies. Ullapool M&VC's exhibition is structured around social themes such as – for the modern period – the Clan System, Townships, Drovers, the Herring, Cleared for Sheep, Ullapool the New Village, Crofting Protected, Emigration. Its narrative mainly hinges on the curator's voice which provides the main storyline with accounts from historical actors featured separately. In the curatorial text, Clearances are presented in general unspecific terms, which might apply to a variety of properties in the Highlands, and reveal little as regards local estates and landholders:

> The estates needed a cash income, much in excess of what the poorer tenants could provide from their rent. Various landowners increased the cattle stocks and then from the 1800s increasingly let the land at a good rent to sheepfarmers …
>
> The sheep required extensive grazings so that the tenants were evicted from their holdings …

Abandoning a wide-ranging perspective, the focus narrows to one particular example from Wester Ross in the section dealing with land legislation. The context leading to reform includes the mention of an incident illustrating the tensions arising from evictions – the well-publicised Leckmelm clearances orchestrated by the new estate owner A.G. Pirie in 1879; such incidents are said to have pricked the conscience of many MPs and resulted in the appointment of the Napier Commission.

The themes listed above show that the division and headings of the exhibition's story-boards underline what was felt were defining moments and issues for the Lochbroom area; in particular the significance of crofting to both the history and the identity of the region today is signalled through narrative structure with a separate space allocated to developments leading to the formation of crofting townships along with the legal protection crofting came to enjoy. It also filters through the 'conversation' taking place between the final commentary of the curator – 'crofting is a traditional way of life in Lochbroom and has given a distinctive stamp to the landscape' – and the contemporary evidence selected; the sources chosen juxtapose a critique of the situation in existence before the Crofters' Act, in the words of a local witness to the Napier Commission, and James Ross' poetic evocation of Murdo Munro's eviction:

> To an old ruined barn with an open door
> Through which the cold snow drift and loud wind did roar
> That brave hearted Gael with his family fled
> Deprived of a home to cover his head
> Rouse clansmen rouse with our voice, with our pen
> Let the echo resound over mountain and glen
> Let us rouse the whole land and demand a death blow
> To such laws as evicted brave Murdo Munro.

It is this emphasis on clearance, land occupation and land reform which, to some extent, sets Ullapool M&VC's exhibition apart from other Clearances narratives concerned with contextualising emigration.

In approach, the storyline shares similarities with many emigration narratives, as not only does it look into destinations, causes of departure and settlers' contribution to their new homelands in general terms, but it also spotlights the voyage of the Hector and the experience of William Michael from Creich who had emigrated to Australia in 1852. Yet, importantly, land access and reform also leak into the substantial section devoted to emigration. As part of the process and motives for emigration is featured the authorities' view of it as a solution to land congestion still favoured at the time of the land agitation, a perspective presented as anathema to the Highland Land League:

> We have thousands of acres of land under sheep and deer which, if properly and economically used, would be an immediate relief to those who have been victims of rack-renting, oppression, and eviction.

Furthermore, the exhibition's overarching focus is 'the story of Lochbroom'. It is brought up to date with modern day fishing and the accompanying video concludes with the trope of struggle against a harsh environment and eventual victory, depicting modern-day Ullapool as a thriving and vibrant community.

Mallaig HC's exhibition, in contrast, focuses on the particular and the structure of its narrative hinges upon geographical locations: Mallaig, Aideagan, Cnoideart, Mòrair, Rudha Arisaig. Meshing curatorial voice and accounts from historical actors, insight is given into the distinctive shape events took locally and the exhibition discriminates between the effects of clearance policies in various parts of West Lochaber highlighting disparities in terms of periodisation, of ownership and sales of estates and of orientations pursued with their ensuing social, economic and demographic effects. Clearances thereby appear as a succession of differentiated episodes with their specific set of circumstances in an attempt to steer clear of oversimplification and generalisation. The personalities and financial standing of families and individual owners also transpire so that proprietors are not all cast as evictors as shown with the example of Eneas Ranald Macdonell said to have 'distinguished himself as a supporter of the local people' in the late 1880s and the 90s. Whilst a high premium is placed on demographic data to convey information as regards the impact of policies – whether it be population change in Knoydart or the proportion of fishermen and shepherds/sheepfarmers in North Morar over the second half of the nineteenth century – the experience of expulsion and its memory rely on contemporary accounts, such as Donald Ross' graphic rendering of the Glengarry evictions or the evidence given before the Deer Forest Commission in 1894. At the same time, excerpts from the Commission are used to showcase a range of perspectives on crofters' conditions subsequent to the Clearances – landholders and various shades of crofters' views. However, if the exhibition strongly foregrounds population decline and redistribution, changes in occupation and conditions of living, neither crofting nor emigration stand out as major themes, but simply form a significant sub-text in the narrative.

Crofting and the Land Wars

In the five remaining venues comprised in this strand, crofting is the crux of narratives which either foreground the Land Wars and the Crofters' Act – the defence or protection of crofting – or the analysis of crofting as a form of agriculture peculiar to the Highlands and Islands. Three are located on Skye – *Aros*, Skye Museum of Island Life (SMoIL) and Colbost Folk Museum (Colbost FM) – one on Harris, *Seallam!* Visitor Centre (*Seallam!* VC), and the last one, the Highland Folk Museum (HFM), in Newtonmore. Different though they are in museological terms (see Chapter 6), it is their approach to clearances through crofting history which enables meaningful comparisons; emphasis will therefore be put on its treatment and representation. It is worth highlighting that, with the exception of *Aros* and *Seallam!* VC, they are folk museums whose main purpose is to exhibit material culture; when historical context is provided, on the one hand it is in distinct spaces, either a separate room or dwelling, on home- or designer-made story-boards, and on the other hand, it is largely restricted to crofting issues. Conversely, *Aros*' exhibition resorted to multi-media technology and was segmented around a succession of tableaux with life-size models, three of which were directly relevant to the period of the Clearances and the Land War: the Skye poetess of the Land War, Mary MacPherson; emigration from Skye with the figures of Samuel Johnson and James Boswell and finally members of the Napier Commission taking evidence locally and listening to a crofter from Braes. Its brief was to tell Skye's story and its thematic content was wider-ranging than the folk museums.

Dominating *Aros*' narrative of the Clearances were the themes of landlords' oppression and people's resistance and resilience; the history of Skye was essentialised around this duality in a presentation which, although using a variety of contemporary evidence, remained unambiguously monolithic. Whilst the opening frame of the exhibition staged Mary MacPherson and the crofters' struggle of the late 1880s, the central scenes focused on the 'injustices' suffered by the people: for instance, the disintegration of the clan system and emergence of rapacious landlordism; 'slavery' with a ship contracted by the tenant of a local laird transporting a kidnapped group of Skye people to America in 1739; and the experience of emigration for local families with the resulting haemorrhage from Skye in the eighteenth and nineteenth centuries. The final sections returned to the Crofters' agitation of the late 1880s with the Napier Commission visiting Skye, to finally conclude on contemporary Skye – modern crofting, fishing, forestry. The sense of wrong pervaded a narrative in which emigration, when linked to clearance, was encapsulated in the 1850s famine evictions from Boreraig and Suishnish recorded by Archibald Geikie. Emigration was associated with exile, distant destinations precluding return, and the dispersal of clans and kin; it denoted unmitigated ordeal foreclosing alternative readings such as the quest for opportunity and a new life abroad. In a similar vein, the section devoted to late nineteenth-century crofters' protest and actions gave much prominence to the experience of imprisonment to which many local activists were subjected, somewhat tempering the sense of achievement and security which the Crofters' Act was meant to epitomise and the final part of the exhibition showcased. Survival thus appeared to have been achieved in the teeth of rigid local and national power and rules. Crofting, we were told, endured 'because of the gains of the crofters in the past', yet at no point in the exhibition were the origins of crofting as a system of land use and way of life defined, and any uninformed visitor might easily have presumed that Improvers had not made it part and parcel of their programme of transformation and that, therefore, it was an 'indigenous' form of agriculture.

In common with *Aros*, the other two Skye museums, Colbost FM and SMoIL deal with the Crofters' War and Act and all three, incidentally, highlight developments at Glendale and Braes as notable acts of resistance at a local and wider level. Both folk museums are highly idiosyncratic and it is worth spending time describing the nature and style of the information that accompany their displays. At Colbost FM, which is a reconstructed nineteenth-century blackhouse with, behind it, the replica of an illicit whisky still, light is shed on the socio-political context of this dwelling, by means of a brief note introducing the unfolding of Skye disturbances in the 1880s and the results of the crofting legislation for the Glendale Community. The note vividly starts with an imaginary re-enactment of a scene witnessed from the vantage point of the house: 'From here in 1882 Thomas Campbell watched a troop ship anchor in the Loch below …' and thereafter sketches, in a succinct paragraph, the developments which followed and led to Glendale community's unique status – until comparatively recently – as a community in full ownership of its land. One striking feature of Colbost FM is that a sense of past events is gathered from a series of now much faded press cuttings pasted on boards, mainly taken from the *Press and Journal* and the *West Highland Free Press* and, as regards the Land Wars and its impact, essentially dating from the late 1970s and 1980s; many were written by James Hunter who, incidentally, was also consulted for *Aros* exhibition.

Figure 4.3 Colbost Folk Museum, Isle of Skye
Source: Laurence Gouriévidis 2007.

The narrative constructed was literally hand-picked by the museum's owner. The tone of the selection on offer is one of celebration of past resistance with a hint of nostalgia as the following titles illustrate 'A bloody confrontation on Skye', 'Bring back the spirit of rebellion', or 'Revolution in the Making'. The selection was a child of its time in the sense that it echoed the historiographical debate which had centred on the alleged passivity of the Highlanders during the Clearances (Hanham 1969b, Richards 1973b, 1974). However its aim was firmly rooted in the present, as the past was not simply conjured up for its own sake, but used as a paradigm for modern concerns.

Present concerns are also the hallmark of the narrative found in the other idiosyncratic venue, SMoIL. Since its opening in 1965, the museum gradually expanded to include seven cottages when last visited in 2007; initially visitors were guided round and told about the historical context before panels were introduced. Historical information as regards crofting and the Clearances period has for some time been located in the 'ceilidh house', and can be gathered from two main thematic home-made story-boards: 'A Short History of Crofting' and 'An Introduction to the Isle of Skye'. Whereas the latter text contains up-to-date information taking the visitors to the present day, the history of crofting was written in the 1980s with references to crofting legislation and developments stopping with the 1976 Crofting Reform Act and remaining silent on later developments, such as community buy-outs and the 2003 Land Reform Act. Textual data and presentation testify to the growth of the displays and to the owner's inclination towards accumulation and layering of information – and objects for that matter – rather than renewal and revision. However, the number of pages and amount of information in the 'ceilidh house' is awesome and many are the visitors who merely graze over the content, picking bits here and there and it is doubtful whether many absorb what are essentially 'books on walls'. Not surprisingly then, of all three Skye venues, the exhibition gives the most extensive account of the Land War and ensuing developments and legislation, including political changes at a national and local level – impact of events in Ireland, successive elections and emergence of newspapers and pressure groups defending the crofters' cause. In form, it resembles a traditional potted history with no voice other than that of the curator/owner.[3] In tone and approach, the narrative differs noticeably from *Aros*' style, as oppression is not overstressed, but the impositions and restrictions inflicted on the local tenantry are used to justify their eventual opposition and action, and more importantly perhaps, the origin of crofting and its development are charted. Crofting is presented as the product of socio-economic change in the region: 'crofts, once laid out to provide homes together with a home food supply, are a survival of past economic conditions'. The existence and continued survival of crofting is ascribed to a set of combined circumstances, not least the conclusions and work of the Crofters' Commission. Clearly the narrative, traditional though it be in format, does not strive for emotional appeal by presenting a deeply fractured vision of the world, but weaves together a complex and changing set of conditions and causal relations.

3 The text of the exhibition was available in pamphlet form in the late 80s and early 90s: J. MacDonald, *A Short History of Crofting in Skye*, n.d.; it was for a time also available at: http://pages.eidosnet.co.uk/~skye/crofting.html [accessed 10 June 2008].

Like *Aros*, *Seallam!* VC's brief is wide-ranging as it presents Hebridean history from pre-historic time, and the Clearances appear in several sections related either to crofting history or emigration. *Seallam!* VC's exhibition and its predecessor, *Co Leis Thu?*, although both dedicated genealogical centres, do not simply aim at contextualising emigration from the outer Hebrides, but cater for a much wider public (see Table P.1 for precision). *Seallam!* VC much more so than *Co Leis Thu?* as it is also child-friendly, offering activities to younger visitors in each section of its display. It has built and expanded on the approach that *Co Leis Thu?* pioneered; both exhibitions give prominence to the history of Harris, but *Co Leis Thu?* laid emphasis on local events and personalities, so that localities and clans or major owners structured its narrative, the Clearances for instance being itemised by area – Horgabost, Sgarasta, Tarasaigh.. Harris remains very much at the heart of *Seallam!* VC's narrative, its past and present directing the themes selected and details or examples highlighted, however characteristic features and changing conditions now organise the narrative as these titles indicate: 'On land and sea', 'Fruit of the loom' – dealing with Harris tweed – or 'They spread across the world to make a new home'. Clearances are introduced when the development of the crofting system is presented and its fate is related to the collapse of the kelp industry, crofting being described 'as a means of gathering as many people as possible on the land to work in the kelp industry'. The 'return' of crofting to the island is later outlined in the final section 'from now on' and is attributed to inter-related factors such as the various government schemes and policies resulting from the 1886 Act, but also the decline of sheep-farming and the land-raids which followed the first world war; crofters' actions are not singled out any more than the other arguments listed.

Importantly, Clearances are also mentioned in connection with specific emigratory waves from the islands. Yet, like Ullapool M&VC, where emigration is also contextualised, and in contrast with other venues where it is the driving force of Clearances narrative, *Seallam!* VC also foregrounds the present fate of Harris, and although its narrative follows emigrants to their new homelands, it returns to and ends on Harris where in-migration is occurring and newcomers are moving to reside in the Hebrides from other parts of the country. Significantly, the child-oriented cartoon illustrating the emigration theme epitomises this approach:

- It seems that lots people have left the islands in the past.
- Yes, but others have come to settle here. There's Gavin, he's moved here from the south …

This short dialogue highlights the prevailing characteristic of an exhibition strongly anchored in the present, depicting Harris as fully engaged with the modern world and with a lively linguistic culture denoted by the wide use of Gaelic in the display. In terms of narrative control, in contrast with the previous exhibition, in which only the curatorial voice appeared and anecdotes were woven within the curatorial text, *Seallam!* VC offers different layers of information both with respect to source and targeted age-groups, as historical actors are heard, featured separately and, uniquely, short imaginary dialogues in the shape of cartoons represent Bill and Chris Lawson answering children's questions.

In keeping with the objective outlined in the 1981 Williams Report, *A Heritage for Scotland*, the HFM in Speyside when it did include boards was intended to:

> help you [the visitor] as you travel through the Highlands and Islands, to understand how more recent features of the agricultural landscape and the pattern of the settlement you see around you have been shaped.

It plotted the emergence of crofting in a narrative couched in broad geographical terms, meant to throw light on the economic, social and political events which had contributed to transforming the Highlands and Islands and producing a distinct form of husbandry – crofting – which, viewed from a long-term perspective, was 'decidedly modern'. It is in the context of the emergence of a new agricultural system that the Clearances were presented as the systematic process of removal 'of entire communities to a new life in an alien and hostile environment on the coast …' on crofts which 'were never intended to enable tenants to be self-sufficient'.

What HFM, *Seallam!* VC and SMoIL underscored and is noticeably absent from venues in which crofting remains undefined, is that it was the product of socio-economic engineering on the part of a landowning class intent on capitalising on both land, alternative industries and labour force. What all three venues also conveyed – to a limited extent at *Seallam!* VC, explicitly and formally at the HFM – is that in spite of its modernity, crofting came to incorporate and be associated with a traditional heritage of practices, customs and values, making it the tangible symbol of much more than just a socio-economic way of life. One striking element made apparent in many of the venues in this strand – even those which do not explicitly engage with the history of crofting per se – is therefore, not only the representation of crofting as the very essence of the region's individuality in physical terms, but also its close interconnectedness with an enduring intangible heritage in the northern districts of Scotland. As museology developed and evolved, this holistic approach to history was to underpin much of the philosophy which directed museological interpretation. However, anyone visiting the HFM in the twenty-first century would not be enlightened about crofting history as contextual information has disappeared.

Narrative strand ii: National

Just as the HFM might be regarded as the interpretive hub of the region's recent past and material culture in the Highland region, so the Museum of Scotland (MoS) and the National Museum of Rural Life (NMoRL), in view of their status as national museums and their location in Scotland's central belt, may legitimately be expected to provide a core narrative covering the whole of Scotland – a central point of reference for visitors who are likely to radiate outwards from the centre. How are the Clearances emplotted in the national narratives of Scotland's past? Neither national museum addresses the Clearances from the perspective of the dominant nineteenth-century elite discourse of Improvement: the intellectual or rational justification for removals. In local museums, 'Improvement' is rarely defined and is frequently revealed implicitly as in venues dealing with crofting agriculture; in others it is introduced by means of local examples of eminent local improvers, the most potent of whom are the Marquess of Stafford (Strathnaver Museum and Timespan) and Sir John Sinclair of Ulbster whose experiments in Caithness are delineated in Gunn HC. Whereas both national museums provide extensive displays on Improvement, neither strongly implicates the Clearances in its ambit.

At the MoS, in the section specifically devoted to this aspect – part of the Enlightenment theme, entitled 'The Spirit of the Age' – if the transformation of Scottish agriculture with the introduction of new farming methods along with the intellectual movement which propelled it is tackled, regional peculiarities are ironed out and little percolates about the shape Improvement took in the Highlands in agricultural or social terms. Brief allusions are made to the Highland and Agricultural Society of Scotland, seen as a crucial platform for disseminating information – 'the main centre for discussion' – and Sheriff Donald MacLeod of Tain. Described as an 'enthusiastic improver of his own estate' who also helped develop local infrastructure, he is one of the examples selected to typify the process, but the focus of the glass case where he is presented underscores household wealth and possessions rather than local socio-economic implications. Comparing lifestyles, the narrative contrasts improvers to ordinary people, indicating that prosperity was slow to reach the lower order of society whose belongings varied little. Yet for the Scottish peasantry in general, improvement would have had a disruptive effect on their lifestyles as they relocated to urban centres or became employed as waged labourers; social dislocation and upheaval do not seep through and the very notion of 'Improvement' is left unchallenged with respect to different social strata. The issue of the Clearances is also certainly evaded: to what extent did the Highlands and the Lowlands differ? How was the policy implemented on the ground? What effect did it have on ordinary Highlanders in view of the clan system, its traditional ethos and the socio-economic organisation on which it was based?

Another powerful illustration of this comes from the material evidence forming the backbone of the section depicting rural life in Scotland, 'Living on the Land', which borrows heavily from the Highlands and Islands – including the northern isles – so much so that, in essence, rural Scotland is overwhelmingly defined by its Highland folk culture. Yet regional distinctiveness and the disparities between Lowland and Highland socio-economic experiences are not interpreted as such. The two are subsumed within the evocation of timeless pastoral scenes – often photographic – linking artefacts with pictorial illustration: an organic folklife preserving age-old practices and techniques (Figure 4.4).

Figure 4.4 Museum of Scotland, Edinburgh
Source: Reg Briers 2009.

In the eighteenth century, the visitor is told, whilst modernity 'gradually affected the way people lived and worked, *in some more remote areas* the use of local materials continued well into the twentieth century' (my emphasis). If the visual collision between Newcomen engine and cruck-framed house is meant to act as the metonymy for technology and rural life and signify disruptive transformation (McKean 2000: 140), not only is little effort made at showing the impact of changes wrought by 'the spirit of the Age' or 'industrialisation' on the Scottish countryside and folks in general, but also no explicit attempt is shown at distinguishing between the effects of such changes on the Lowlands and the Highlands as regards chronology, form and consequences. Neither the Improvement nor the rural sections of the exhibition have anything to say about the Clearances period, mainly because the dominant tropes of these parts of the exhibition as regards the peasantry are continuity and uniformity of practices associated with remoteness.

Whereas changes in national economics and politics are largely inherent to local museums' narratives and were at the heart of the short-lived exhibition in HFM, seemingly the MoS has chosen to 'evict' the Clearances from its own narrative when engaging with the Age of Improvement and rural life in Scotland; so where are the Clearances featured in the MoS' rendering of Scotland's national past? They are associated with Scots on the move or emigration in the section entitled

'Scotland and the world' which traces the movement of Scots to various parts of the world – mainly the British Empire. Significantly the pictorial illustration on the wall surrounding the video which contextualises Scottish emigration throughout the ages, is taken from the pamphlet *Real Scottish Grievances* by Donald Ross and features the scene of an eviction. The way in which the MoS addresses the question of the Clearances will therefore be contrasted to other museums where narratives foreground the movement of Highlanders and Islanders abroad – in the emigration strand.

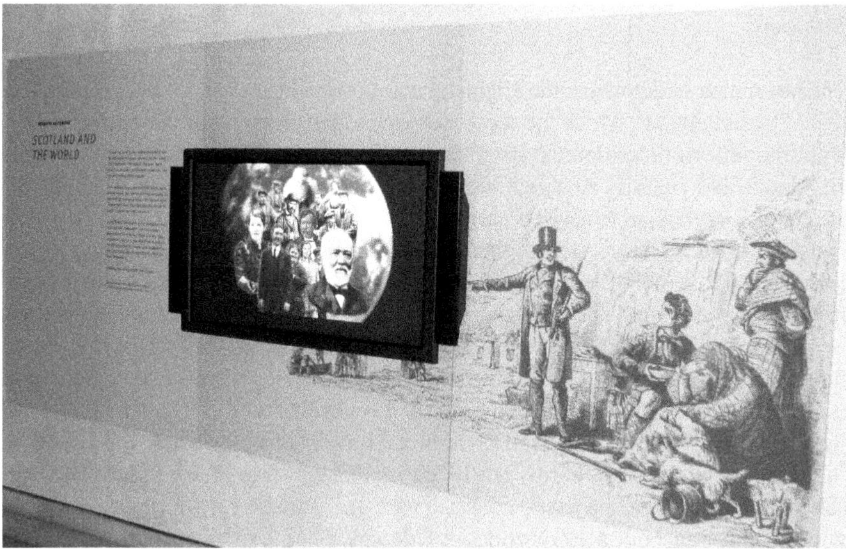

Figure 4.5 Museum of Scotland, Edinburgh
Source: Reg Briers 2009.

Both national museums chose to inscribe their interpretations of the Clearances in narratives underlining the experience of the people, with particular regard to displacement, uprooting and landlordism as a socially disruptive force, rather than the rationale framing changes in land use. At the NMoRL, whose exhibition was conceived around a tripartite thematic structure – land, people and tools – the Clearances are presented in the people or social history section, whilst Improvement is introduced earlier, when the exhibition portrays the transformation of the landscape. There, few of the examples linked with the introduction of new agricultural practices and methods refer to the Highlands as the presentation, perhaps understandably, is clearly geared towards providing contextual information for the surrounding working farm.

When the museum was first opened, the Clearances display referred exclusively to the Highlands and Islands and, in terms of communication, was based on illustrated text panels; five years later, the focus was altered and the term 'Clearances' was widened to assimilate the Lowland experience;[4] an audio-visual display was also added. In the words of the introduction to text panels:

> The Highland Clearances are remembered as a time of great hardships. Were you aware that the Lowlands were also cleared of people? Prior to the Highland Clearances, thousands of people were removed from Lowland farms and resettled elsewhere. We have, until recently, tended to forget the Lowland Clearances story.

The key notion under which the Highland and Lowland experiences are subsumed is that of 'resettlement', which the accompanying video clearly articulates concluding that: 'the period of resettlement [from the Lowland migration to the clearances in the Highlands and Islands] changed Scottish life forever.' Yet, although 'Clearances' as a term is reclaimed to serve a national narrative, as a process, to some degree it is differentiated. The disparities that agrarian changes entailed at a regional level are outlined in terms of land use, forms of pastoral farming and, to a lesser extent, resettlement provision and opportunities. The chronology of their implementation is clearly demarcated, as are such social conditions as the effects of the potato famine in the Highlands. More surprisingly, instances of resistance to changes and protest are produced in relation to both the Lowlands and the Highlands, making light of the fact that the 'Lowland Clearances' have been described by historians as 'a silent revolution' with 'hardly any evidence of angry protest or collective unrest' (Devine 1999b: 150), in stark contrast with the situation in northern Scotland which witnessed both acts of collective defiance and expressions of opposition throughout the period (Devine 1989: 148-68), culminating with the Crofters' War of the 1880s. Curiously, in the display, whereas Lowland resistance is illustrated with the Levellers' destruction of stone dykes in the early 1720s, for the Highlands and Islands, the Land War is not featured, at least with examples pre-dating the 1886 Crofters' Act. Among the two artefacts used is a 'rabhadh' or warning served in 1888 at Aignis on Lewis during one of the land raids. To be fair, another part of the 'people's' exhibition entitled 'Church, school and politics', does have one exhibit whose label mentions crofters' actions in the North West Highlands prior to government legislation, but it is reduced to one diminutive instance of people's challenge to authority and elite power, and whether visitors will associate it with the much broader movement of rural protest in the Highlands or remember the allusion when moving to the Clearances section is extremely doubtful.

Deciphering the Clearances as an experience uniting all Scots is not unproblematic and leads to lack of clarity. Ambiguities come to light in the museum's treatment of the potato famine period which silences cultural fissures

4 Personal correspondence, 27 February 2009.

within Scotland conjuring up the vision of a united front in the face of calamity. Whilst the attitude of Highland landlords is conveyed with modulation and all are not tarred with the same brush, a 'minority' said to have provided assistance, reactions from the 'Lowlands' – whatever the term refers to – are not as nuanced. Starvation is said to have been 'narrowly avoided by help from the Lowlands', and indeed there was substantial help from the region, not least from the Free Church, yet the Highlanders were also disparaged and disdained in the racial discourse peddled by some contemporary Lowland papers. The approach also leads to a lack of socio-cultural and ethnological specificity. This is most significant with regard to the consequences of Clearances in the Highlands. Because the clan structure is not featured, neither are the mental remnants of former clan organisation conditioning much of the people's reaction to evictions. More strikingly still, although the term 'crofts' is used, coupled with the fact that they were not 'self-sufficient', crofting is left to speak for itself; its roots are not explicitly specified and its long-lasting legacy is consequently screened out.

In the final analysis, the narrative features an interesting reversal of more conventional interpretations of the Clearances as a process, which emphasise the widespread social effects of agricultural transformation propelled by the industrial revolution before narrowing their focus to regional peculiarities, such as Clearances in the Highlands and Islands. Here the marker 'Clearances' is used as a starting point of collective understanding to introduce the story of sweeping changes leading to Scottish urbanisation and the inadequacy of towns and cities in assimilating the rate of internal migration. As a notion 'Clearance', in the display, connotes not only 'an emptying land', but the destruction of a traditional way of life, in broad social terms, with its familiar landmarks. As opposed to the MoS, where the Clearances formed one regional example in the wider story of Scottish emigration, at the NMoRL, they have become a major social theme read in terms of its national resonance.

Narrative strand iii: Emigratory

In a number of venues, emigration and overseas connections are dominating paradigms in Clearance narratives, partly owing to the nature and purpose of these museums, partly as a result of interpretive choice. Apart from the MoS, where the experience of emigration epitomises the scattering of cleared communities throughout the British Empire, Clan Museums – the MoI and Gunn HC – underscore the end of the clan system and the dispersal of the 'clan' abroad; Glengarry HC was to some extent established to respond to demands from descendants of emigrants, and at Arran HM, the three-part display entitled 'Clearances', is segmented around removal from the island, voyage and settlement in Canada. In these narratives, the Clearances – substantial though their part might well be in the storyline – are often used as a means of contextualising departure, reasons for leaving and life in new surroundings, and are subordinate to this main overarching theme. To analyse the treatment of the Clearances and their emplotment in those narratives, this section will be articulated around two issues: the Clearances as contextual background and the assessment presented of the nexus Clearances and emigration; it will compare the different approaches adopted and their effects.

The Clearances as contextual background

Renderings of the Clearances depend greatly on the way narratives are configured, some are partitioned around changes in estate ownership as at Glengarry HC, others around events, such as the '45, as in Clan museums, so that the periodisation constructed may not foreground the Clearances, however, they always encapsulate the move from clanship to commercialism. The compact display at Arran HM is one which details the shape of changes on the island documenting the economic conditions – lack of alternative employment – which led to the departure of islanders, particularly those leaving on the *Caledonia* in 1829 to start a new life in Canada, an emigration organised by the estate and partly financed by the proprietor, the Duke of Hamilton.

Figure 4.6 Isle of Arran Heritage Museum, Brodick
Source: Laurence Gouriévidis 2007.

The curatorial text emphasises the paucity of records available as regards population displacement and no eviction scene is featured. Similarly no eviction scene is featured at Glengarry HC where, unlike the exhibition at Mallaig HC – which deals in part with similar estates – visitors will find no account of the widely reported famine evictions in Knoydart, on the remnants of the MacDonells of Glengarry's Highland estate. However, they will learn of the critique levelled against them in another part of the display featuring a neighbouring owner: the MP Edward Ellice, who denounced the ultimate purpose of those clearances meant to divest the estate of its pauperised population and avoid assistance in famine relief. His property, when he acquired it in 1839, was described as 'hopelessly insolvent' and its inhabitants 'in a most wretched conditions'. Presented in giant book form in a corner of the research room and organised around the successive owners and division of the Glengarry estate, the nub of the narrative is on the close interconnectedness between clearances and emigration, so that little transpires about individual removals and the conditions of local inhabitants beyond these brief allusions. As with Strathnaver Museum however, the visitor at Glengarry HC is encouraged to visit the remains of a settlement, Daingean, cleared in the late eighteenth century and opened in 2004 (see next chapter).

The two Clan museums approach Clearances very differently. Gunn HC is most uncharacteristic as it uses the writings of one of the most famous sons of the clan: the novelist Neil M. Gunn whose novels provide a separate, imaginative and often emotional commentary in each part of the display. In the section 'the '45 and the Clearances', removals are ones of the forces listed, alongside 'poverty' and 'sheep', to account for the large number of emigrants driven from the region in the early nineteenth century. The narrative homes in on the Kildonan evictions on the Sutherland estate and the opposition mounted by the people, many 'men of Clan Gunn' against the Staffords' plans and factors. The words of Donald MacLeod's *Gloomy Memories* and excerpts from *Butcher's Broom* bring the process of eviction to life and bespeak its mental scars:

> EVICTIONS
>
> And eviction was a terrible thing, for it meant the tearing up of life by the roots and the throwing of it on a sea-bench or on a foreign strand where everything that made life comely and happy and to be desired would forever be denied …
>
> THE CLEARANCES – a betrayal
>
> … We did not blame them for murdering the body; we blamed them for murdering the soul; we blamed them for taking an order of mankind, faithful and loyal, who in the courses of ages had given light to the world … great music and story; for having taken that order to which they themselves belonged, and for 30 fleeces of wool to have betrayed it in its own garden and destroyed it (Both excerpts from Neil M. Gunn, *Butcher's Broom*).

Whether the emigrations which followed, were enforced by the estate or in reaction and rejection to the changes imposed, the narrative does not say and, indeed, the issue pales into insignificance beside the psychological and social trauma experienced by the evictees.

Figure 4.7 Museum of the Isles, Isle of Skye
Source: Laurence Gouriévidis 2007.

By contrast the MoI, which provides the most extensive treatment of the Clearances in museums within this strand, does not include vivid representations of the evictions implemented on the various estates owned by the descendants of the chiefs of Clan Donald and related branches. For the Clearances period, the narrative is split between two rooms, 'Chiefs to Landlords' and 'Clansman to Crofter', a separation vividly materialising the move from kinship to class division and social dislocation, a separation also emphasising the extent to which the final room 'emigration' trailing the clan around the world as in a global family search, is a modern cultural construct, yet with powerful attraction for those with similar patronymics – and rewarding prospects for the tourist industry (see Chapter 7). Interestingly, the transformation experienced by the two social groups outlined is styled very differently. On the one hand, the combined effects of the changes affecting the chiefs' way of life, such as increased spending and residence outside the region, coupled with deteriorating economic circumstances is charted through individual examples of sales of estates – Clanranald who, 'by 1838 had nothing left except Castle Tiorram, the clan's ancient stronghold', Glengarry who sold 'Knoydart, his last remaining land, in 1853', and Lord Macdonald who 'sold North Uist and the northern part of his Skye estates in the 1850s.' On the other hand, the picture painted of the consequences on 'clansmen and women' of the agricultural transformation of estates is curiously de-personalised (Figure 4.7).

No specific eviction is recounted and whilst crofting agriculture is defined, whilst the disastrous fusion of sub-division of holdings, population growth and reliance on potato cultivation is conveyed, whilst the cultural impact of clearance and exile transmitted through Gaelic oral culture is described, no pinpointed information is imparted as regards Skye or other parts of the Highlands and Islands belonging to leaders related to Clan Donald. The curatorial text moves to the impersonal and generic, as shown in this portion devoted to the clearances:

> The period from the late 18th century to the first Crofting Act in 1886 is known as one of clearance. This process of dispossession was not just the large-scale eviction of whole townships. It was the relentless trickle of people away from their land when faced with rising rents they could not pay. This removal of the lesser folk from the land went on all over Scotland. It went on all over Britain at the time. But it was particularly traumatic to the Gaels because it went against the old concept of duthchas, with its inalienable link to the land and the constant protection of the chief (…)
>
> The clearances bled the Highlands of its people. Some left for the industrial cities of the south. Others ventured much further afield. They were the Emigrants.

What occurred locally is concealed under the mantle of universality of experience; the words 'clearances bleeding' the region remove the notion of individualised responsibility and add that of the inevitability of a process in which expulsive and oppressive forces are not personified – in stark contrast with the previous exhibition at *Aros* where the well-publicised famine evictions on the Macdonald estate were earmarked. As implied in the final sentence, the main thrust of the narrative is not the locality nor its identity, but the outward movement of clans people originally scattered in the Highlands and Islands.

Nowhere is the use of Clearances as a structural link leading to the representation of emigration more conspicuous than at the MoS, and it makes for a treatment of the Clearances riddled with ambiguity and haziness. For a start information is fragmented and has to be pieced together from various sources and areas: firstly from a label commenting on the painting by Sir Edwin Landseer, *Monarch of the Glen*, displayed at the far end of the 'Scotland and the world' zone; secondly, from the video setting the context for the emigration display; and finally from a glass-case summing up the causes of emigration – 'reasons for leaving'. The painting is used to signify the discrepancy which existed between artistic representations of the Highlands, essentialised through 'romantic wilderness', and the grim reality of transformation of land use and social relations with its trail of misfortune and hardship for those cleared from the land, 'some forcibly removed to coastal settlements or to find a new life overseas'; it is also conveniently used to shift the narrative's focus to emigration. In the video which follows, the nineteenth century, marking the beginning of 'mass exodus' from Scotland, opens with the Clearances, epitomised by the Sutherland estate and supported by startling images: homes ablaze and the title page of the 1853 pamphlet *Extermination of Scottish Peasantry* by R. Alister. The voice over observes that 'forced emigration changed the mood of the country, Scotland's painters and poets adopt[ing] themes of sadness, loss and separation'. Finally the glass-case follows on from the outline sketched in the video and Clearances are correlated with forced departure in a paragraph starting with the words: 'not all emigrants left voluntarily'. There lies the most strident flaw in the MoS' rendering of a process, which lasted over a century, and is made to stand as the quintessential symbol of coercive emigration, failing to show the many shades of experience which accompanied the different emigratory waves linked with clearance. As a result, it overlooks those who left voluntarily in defiance of plans either suspected or thrust upon them, and the many estates, including the Sutherland property, whose initial intention was not to associate clearances with emigration, but with retention of population through resettlement in coastal villages where alternative industries were to develop – a much needed labour force (Richards 1982: 284-95, 2000: 122, 127, Devine 1994a: 37). Ironically, the MoS' narrative manufactures its own oversimplified and archetypal vision of the Clearances, yet the artefacts exhibited alongside 'reasons for leaving' might have told an alternative story, had they been interpreted differently. Of the eleven items displayed, some are very loosely connected with emigration but are indicative of the so-called 'improving' economic rationale which underpinned the Clearances, of internal population re-distribution, of the redeployment of an agricultural people to occupations alien to them, and of the reactive aftermath of the policy and events in the region – the Land Wars (Figure 4.8).

The silver teapot belonging to Rev. Donald Sage, the Sutherland Minister who recorded the Kildonan and Strathnaver evictions in his memoirs *Memorabilia Domestica*, the silver teapot rewarding the best specimen of kelp in Ross-shire and Inverness-shire in 1821-1822, the Duke of Sutherland commemorative medal on the occasion of his death, the picture of Dunrobin Castle and finally the silver fob watch chain presented by Sheriff Ivory to a policeman involved in the arrest of Norman Stewart, the 'Parnell of Valtos' in December 1884 were all invaluable artefacts for the 'story' of the clearances and their impact on the region and on the population who stayed. Curiously, these items are framed within a narrative of national emigration which, even then, does not convey the complexities of the relation between clearance and departure, let alone those of Highland and Island emigration.

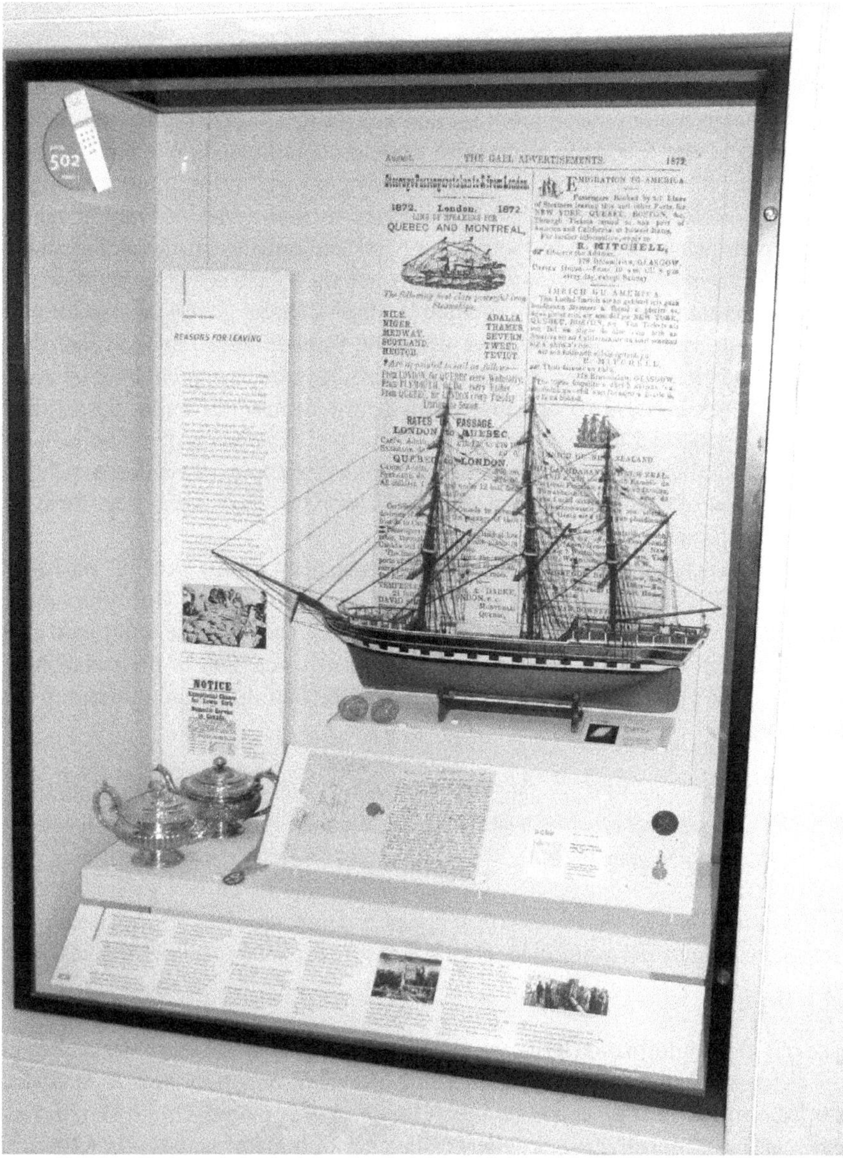

Figure 4.8 Museum of Scotland, Edinburgh
Source: Reg Briers 2009.

The nexus clearance and emigration

Part of the difficulties in weaving Clearances within narratives of emigration is that, for displaced or evicted families, motives for leaving the region varied over the period as the economic and social situation deteriorated. When emigratory phases are not sketched out and when clearance and emigration are simply aligned with force, narratives do a disservice to those who actively sought a new life elsewhere free from the pressures of landlordism. Furthermore, such a broad and oversimplifying perspective does not clarify the social implications of emigration as a policy, variously assisted by proprietors or government, a form of social engineering at a time of crisis when the Scottish Poor Law was intensely debated (Devine 1988, 1992: 84-103). The MoS does not cast much light on those issues, mainly because its narrative shifts to the fortunes of exiles abroad and becomes segmented along destinations where the experience of individuals or groups is highlighted. Understandably, few of the examples presented enrich the interpretation of Clearances as the narrative's scope sweeps over a large array of *Scottish* emigrants over a wide timescale.

Also structured around emigrants' destinations, and with a narrower geographical focus and timeline, the emigration section of the MoI offers more information, and does emphasise that emigration could also be a forward and liberating step during the Clearances period, as its discussion of the melancholy of exile in the Canadian boat song points out, in the introduction to this part of the exhibition:

> It gives one view of the emigrant experience. It is both yearning and regretful. We like to believe this nostalgia was typical. Yet it does not represent the whole compass of emigrant life.
>
> Yes, many were forced to leave their homes, and went unwillingly … but many left with hope and expectation. They looked forward to owning their own land …

The rest of the information has to be assembled from piecemeal examples related to the different locations where emigrants built a new life, and from the individualised stories – to which this section reverts – found in drawers which the visitor may pull out. Consequently, it is a little challenging to build a picture of emigratory phases during the Clearances period. For example, although emigration assisted by the Highlands and Islands Emigration Society during the famine, is mentioned in connection with Lord Macdonald's estate in the Australian part, clearances are not specified. The emigratory scheme is disconnected from the issue of eviction and ultimately coercion, both conceptually and physically as clearance and emigration are featured in separate rooms. So the overall narrative at the MoI conveys a sense of continuous departure and of a range of motives and experiences but mainly foregrounds the twin ideas of movement – from the region to different parts of the British empire and from colony to colony – and dispersal over time and space.

Smaller museums, with a more limited geographical remit are more successful in plotting out distinct phases and conditions. As already mentioned, Arran HM spotlights the 1829 estate-assisted emigration providing details of its practical organisation, using the correspondence between the Duke's agent and colonial office. At Glengarry HC, successive waves of emigration with their specific set of circumstances form the core of a part entitled 'Glengarry emigrations', which is chronologically organised. It starts with the move of tacksmen and their followers to New York and, at the time of the American War of Independence, the flight of those remaining Loyal to the British Crown to Canada where they established Glengarry County; continues with the departure of Highlanders from Glengarry or neighbouring areas to Canada between 1785 and 1793, mostly as a result of evictions, rent increase and falling economic prospects; next with the case of the Glengarry Fencibles raised during the Napoleonic Wars and disbanded in 1802, also leaving Scotland because of deteriorating conditions; then the government-sponsored emigration in 1815, of which Glenelg inhabitants took advantage; to finally ends with the 1850s famine-driven emigrations. This last wave, as already hinted at, is quickly brushed over and certainly evades much of the pathos and contention which contemporary records of these eviction-related departures evoke. Another two museums where emigration is a significant aspect of their narratives – *Seallam!* VC and Ullapool M&VC – also differentiate between phases. At Ullapool M&VC two main periods are delineated: one mingling population increase, unreliability of herring fishing and pressure from clearance when emigration became 'a major business', another between 1846 and 1855 when 'poverty and distress reached crisis proportion (... and) emigration was seen by some as one of the solutions to the intolerable conditions prevailing in the north west Highlands', an emigration which could involve public assistance. *Seallam!* VC discriminates between three phases; the first one before the 1820s, when emigrants were stimulated by a 'pioneering spirit'; the second, from the 1820s, when 'sheep was the driving force' and finally the last one, when 'potatoes fail[ed] the people' and emigration was backed by government-sponsored schemes. To some extent, it seems that the local angle, because of its sharper focus, offers more scope for less essentialist and ambiguous narratives, however other factors often modulate this assessment as will be shown in the following chapters.

The impact of clearing policies and the social structure which buttressed them on the *mentalités* and psychology of emigrants is an element which is seldom developed in narratives, beyond brief allusions to landowning aspirations or the nostalgia and longing for one's native place. At Arran HM however, it is a strong feature mainly conveyed through the display of contemporary sources, namely a long poem in which the return to Arran of an old man cleared from Glen Sannox in 1829 is imagined by one of his descendants, *Donald Valdy returns to Arran*, and, above all, an emigrant's letter to an Arran relative written in October 1834 and translated from Gaelic. In it, his appraisal of the benefits of life in a new homeland is clearly derived from the experience of clearance:

> … if they have plenty now it was made by working very hard for it sometimes, but there is one thing that people here know and that is what they have will not be taken away from them.

The museum's narrative of emigration projects a mitigated assessment which might well only be best conveyed, when leaking out of contemporary sources from ordinary emigrants. Other museums adopt a much more conventional narrative approach by singling out remarkable individuals who, in the words of the curatorial text at Gunn HC, 'made important contributions to emerging nations', a statement exemplified by many (male) figures among whom, for Australia:

> the Honourable John Alexander Gunn, a member of the Upper House of the New South Wales parliament, [who] developed a single dose of Anthrax vaccine which was to virtually save the Australian livestock industry.

Certainly the theme of emigrants' contribution to the development of their new homeland is one which dominates many emigration narratives, at times tainted with celebratory overtones, or even verging on self-congratulation when the movement is construed in terms of the national contribution to the New World. Ullapool M&VC is a case in point:

> The Scot emigrants made their mark around the world in Canada, Australia and New Zealand. They played a leading part in the economic and political life of the country and were pre-eminent in trade, banking, engineering and agriculture. A substantial part of the financial investment in the colonies came from Scottish sources.
>
> The long-term legacy of emigration has been a worldwide respect for Scotland and the Scots and a body of people who are proud of their Scottish ancestry.

'Homecoming year' celebrated in 2009, might very well amplify this Scottish view of a process which, in the eyes of native inhabitants of colonised territories, was not one of progress, but also entailed tensions and conflicts, and indeed dispossession and dislocation.

* * * *

Concluding comments on Clearance narratives bring to light two main concerns: one related to availability and location of information and the other to a range of issues raised by the approaches chosen. With respect to information, even if, as suggested in the Williams Report (1981: 14, 16), national museums are meant to act as hubs in a network of localised and specialist satellites with more in-depth exhibitions, the strategy is not entirely successful. The Clearances certainly present difficulties for any single venue because of their fragmented and extremely disparate character – disparities between estates in terms of scale, chronology, intention, effects to name but a few – and museum exhibitions are greatly hampered by lack of space, which limits more exhaustive and qualified analyses, perhaps even more so in national museums. However space alone does not account for interpretive choices and planning. At a national level, the fact that the MoS fails to fully engage with the issue is to be deplored, particularly given the way in which Clearances and emigration are conflated, ultimately isolating clearances from broader questions related to economics or social policies. As a result, the museum does not succeed in providing a broad canvas for a more detailed appreciation of complex features. Pursuing the visual metaphor, the overall effect is more impressionistic than naturalistic and the visitors will depart with little understanding of the period as regards socio-economic and cultural specificities. Even the treatment of emigration is lacking, giving little sense of the disparities between clearances emigratory waves particularly in relation to causes. Worse still, to a considerable extent it reinforces the main stereotypes associated with evictions as the Sutherland estate recurrently encapsulates the process. In the words of one of the Highland museum curators answering the survey question related to the importance of the events to local identity: 'Most parts of the Highlands saw some form of clearance, but most people are only aware of the "headline" events.' Visitors to the MoS will leave with a paltry and unqualified view of the Clearances and simply remember the 'headlines'. At the NMoRL, the narrative's pitch of a national story approached from a social perspective, leaves much of the regional specificity of the process and its consequences undisclosed. Consequently, at a regional level, it seems odd and strangely lacking that no museum should provide a systematic display on crofting and its background, inclusive of the history of the Clearances. Since the HFM discontinued its historical contextualisation, it seems that visitors have to travel much further afield to encounter some form of representation of the events and period, however, in small individual venues anchored in the specific and the local, the provision of an overview of the process as a whole often clashes with their restricted geographical remit and focus.

The choice of themes and approaches in Clearance narratives raises a number of questions. Some are linked with the difficulties of striking a balance between local stories or examples of painful uprooting or resistance and general historical context of social migration and economic change, others with the cloudiness with which the origins of crofting or the causes of emigration are portrayed. Others still are associated with museographical decisions, such as the place and significance afforded artefacts within narratives as well as the meaning they are assigned. Furthermore, the notion of 'contest' so strongly part of the legacy of the period may not necessarily be easily conveyed beyond polarised displays split between the voices of divergent witnesses, and curatorial input may not impart to visitors the symbolic and politically-charged nature of the representation of events so firmly rooted in collective discourses. The purpose of next chapter is to explore further such curatorial and interpretive issues.

Chapter 5
Museum Interpretations of the Clearances: Issues and Strategies

> *Some historians have argued that the clearances have been manipulated by underplaying what would have been the results had the communities not been cleared from the land in terms of over intensive farming and the resultant disease and famine. The morality of the present must always impinge on the past in anything but a pure academic situation. No late twentieth century museum tries to survive on academic research alone. However, the survival chance of a museum telling the clearances story without discussing the moral issues would be slim. The creation of a myth can be for positive reasons* (G. Watson 1995: 261).[1]

The narration of the Clearances in the diverse venues surveyed is not only revealing of issues pertaining to the interpretation of the period itself, but also to the representation of history in museums and the means used in order to impart some understanding of the past to visitors with varied backgrounds, expectations and prior knowledge. These issues may be approached in the light of their relationship with history as a discipline and Clearances historiography more specifically, their relationship to contemporary questions and situations in the Highland region and beyond, and finally their relationship to the preservation of heritage and material traces of the past. The analysis in this chapter will be based on the study of survey answers and the communication strategies underpinning Clearance narratives.

History and Clearance narratives in museums

The approach adopted at the NMoRL where the 'Clearances' as a notion are given a national constituency encompassing both 'Lowland' and 'Highland' Clearances may very well be due to historiographical developments (Cowan 1992: 61-83, Devine 1994b: 111-64, 1999b: 147-51, Aitchison and Cassell 2003, Devine 2006); the same is true of the growing preponderance of highland and island emigrants in museum narratives (Harper and Vance 1999, Harper 2003). However if museum interpretations of the past are, to some degree, influenced by the advance and modulations of historical research, museums are seldom concerned with conveying a sense of historiographical evolution or debates, but with imparting to a varied

1 Graeme Watson used to manage the Heritage Section of Ross and Cromarty District Council.

audience a sense of time and place, frequently through the representation of direct human experiences. This, in the case of the Clearances, is also refracted through their legacy and its bearing on the present, given, on the one hand, the nature of modern-day Highlands and Islands, and on the other, their recurrent and, so far, unflagging use as a powerful sign in diverse discourses. The extent to which curators exploited or were challenged by this cultural baggage, transpires in the answers to some of the survey questions relating to the interpretation of the period (see Appendix, section 11).

Two related questions were meant to engage with issues of use of sources and viewpoints. The first one focused on whether a particular perspective was adopted in the displays, the second on the amount of consideration modern history writing received – both were implicitly related to the polarised deciphering of sources and construction of Clearances historiography. Answers to these two questions are revealing of internalised perceptions of the nature of history as a discipline, and tacit practices as to the reconstruction of the past in local museums – many of them run by non-specialists. Regarding perspective, out of ten answers, six ticked the crofting community box: SMoIL, Arran HM, Gunn HC, Timespan, *Seallam!* VC, Ullapool M&VC. Two respondents added comments to justify this choice, Arran HM mentioning that 'most material available was from the islanders' viewpoint' and Ullapool M&VC that the slant was 'slight and not overplayed as it was not stated openly in the interpretation text *but implied through the contemporary sources used*' (my emphasis). Glengarry HC may also be aligned with this group as the answer indicated that 'the aim was to allow the material to speak for itself, but it could be said to lean towards the emigrants' viewpoint'. Both Strathnaver Museum's and Mallaig HC's respondents opted for what might best be described as assumed neutrality, the Strathnaver informant commenting: 'we have tried not to take sides' whereas that of Mallaig: 'both [Improvers', crofting community's viewpoints], to try and present an accurate image of social and economic pressures at the time.' The MoI provided no answer.

With respect to whether modern history writing was accorded much attention, four of the informants seemed to place little weight on the influence or use of modern production, implicitly hinting at the authority of local archival material, and, even then, leaving aside the process of selection, although, in fairness, it was mentioned in response to later questions.

> *MoI*: Provides other sources of info/studies on additional areas and material.
>
> *Arran HM*: Very little.
>
> *Strathnaver Museum*: Who are we to argue with modern historians? None of those involved in the museum are historians and are only too glad to make use of the research which they do.
>
> *Gunn HC*: Some.

Timespan: Researchers felt a necessity to be aware of modern interpretation.

Seallam! VC: Most of the information came from Bill [Lawson]'s research in the Hebrides and in Canada, Australia and the US.

Mallaig HC: Negligible: at the time of mounting the exhibition no new factors were known of that affected the Knoydart situation.

Ullapool M&VC: I doubt much. Before my time …

Glengarry HC: Marianna MacLean's *People of Glengarry* was a favoured source, largely owing to her depth of research and lack of judgmentalism. Other sources were Eric Richards and some 19thC writers.

Few, apart from the allusion to 'judgmentalism' from the respondent for Glengarry HC, disclose a sense of the plurality of historical voices and genres, and the successive methodologies, debates and points at issue which have propelled historical research. The issue of use of sources – modern or contemporary – voice and authority relates closely to the changing role of museums in society, a role interrogated by the New Museology and one which sees museums, whatever their type or size, poised between institutions implicated in conferring legitimacy to versions of the past and agencies promoting social strength, cultural understanding alongside local participation and expression.[2]

Local expression undoubtedly leaks through the construction of Clearance narratives, probably more so than some respondents cared to admit, particularly through the curatorial and interpretive strategies adopted. The interface between the curatorial voice and the voices seeping through contemporary sources – text and visual aids – is instrumental in conveying historical understanding as it modulates accounts in ways that are meant to impact on visitors' perception of the past. The manner in which these voices are structured within a text and visually, the way in which they interlock, complement each other or clash, and the way contemporary voices are appropriated by a narrator-cum-curator or isolated from curatorial narratives show the extent to which museum exhibitions may act as virtual forums between a variety of standpoints and experiences.[3] In their construction, displays may make room for dialogue between distinctly expressed texts and open up alternative positions, alternatively they may exclude multiple viewpoints. It is not infrequent for exhibitions to be spatially divided into separate zones: division between context and artefacts or tableaux, divisions between curatorial text providing a general background and contemporary sources exemplifying developments on the ground in specific locations, however these divisions might not necessarily convey a

2 See Chapter 1.

3 Much of this part has been informed by conversations with Mary Hutchison, Research School of Humanities, Australian National University.

multiplicity of perspectives, but buttress a monochromatic picture, reducing and even entrapping the visitor's field of vision.

The vast majority of exhibitions surveyed hinges on conventional forms of museum story-telling with an impersonal narrator-cum-curator, an institutional voice, controlling the narrative, frequently providing causal links and, at times, commenting on developments. Some appraise events, situations and people often relying on authorial irony; at Timespan, for instance, the actions of the owners were sneered at by the narrator:

> The displaced tenantry would move to the seaboard with the opportunity to resettle on allocated crofts lots and start a new life fishing ... The Marquess combined self-interest with far-sighted providence for his wife's tenants. None of whom ever seemed to have uttered a word of thanks in return!

In other instances, the period is articulated in terms of social polarisation with tenants' experiences juxtaposed with landlords' objectives, thus charting the clearances as a contested issue dividing local society along class lines; at Strathnaver Museum – where irony is also a feature – the concluding words of the binary presentation alluded to in the previous chapter, provide the visitor with an unequivocal assessment of the period: 'Many factors combined to make the 1820s a disaster' (Figure 5.1).

More generally, the choice of sources displayed gives texture to the storyline and directs its reception. A majority of venues incorporate the voices of historical actors, equated with the voice of witnesses, quoting or displaying excerpts from contemporary sources. The intention is to present the visitor with veritable slices of the past seemingly unmediated and full of conviction, making the narrative at the same time more realistic and persuasive. Many records have survived, often based on personal reminiscences; they may be the testimonies of evictees, landlords and factors, or simply eye-witnesses such as travellers or journalists, and may take the form of pamphlets, newspapers' articles, emigrants' letters, essays or evidence given to the successive parliamentary commissions dealing with the crofting districts, not least the Napier Commission. Contemporary testimonies are frequently ideological and emotional, and follow a variety of discursive formats closely linked with a specific context. The choice is vast and museum space being limited, it can only be honed to serve the effect required. The act of bearing witness, thereby re-enacted in the museum context seldom stands in pure innocence precisely because it is predicated upon the drastic selection of voices. Without critical distance and clarification or divorced from discordant and contradictory stances, it risks becoming manipulative using 'the voice of the past' like 'the voice of truth' to sanction one particular viewpoint. The underlying selection may tilt the interpretation towards apparent and deceptive polarity as at Strathnaver Museum or, as the respondent for Ullapool M&VC commented, may sway it towards the tenants' perspective because of the emotive content of the extracts. At Ullapool M&VC, the exhibition is effectively organised around separate narrative zones, those narrated by the curator and those given over to contemporary witnesses (Figures 5.2 and 5.3).

Figure 5.1 Strathnaver Museum, Bettyhill, Caithness
Source: Laurence Gouriévidis 2007.

At Mallaig HC, where the curatorial voice is muted, multiple excerpts or quotes are used, shattering the view of communities, be they landowning or crofting, speaking with a single voice; this makes for a more confused but also less essentialist picture. At any rate, the overwhelming use of dissonant voices does convey the sense of an imploding and fragmenting society and the strain of social relations along with, in some cases, a sense of the anxiety of human condition. This certainly reflects a perspective which steers clear of re-creating a master narrative slanted towards the elite, where alternative positions are muzzled and the people's version does not percolate. A substantial majority of Scottish museums in fact, including the nationals, portray the tearing apart of the social fabric and the human experience lying behind socio-economic policies and changes, thus foregrounding the conditions of the displaced, the uprooted and the emigrants – the people's experience. One venue, *Aros*, emphatically foreclosed alternative perspectives with one and only viewpoint – be it expressed by the narrator or contemporary witnesses.

Figures 5.2 and 5.3 Ullapool Museum and Visitor Centre, Wester Ross
Source: Laurence Gouriévidis 2008.

In this context, a popular curatorial device used to communicate a sense of the past hinges on the principle of vicarious empathy conveyed through imagination and individualised experience. At its most intrusive, the narrative voice may directly address visitors inviting them to identify with a particular situation or individuals, eliciting empathy with the experience of historical actors. Two examples stand out in the museums surveyed: at Strathnaver Museum, the visitor is requested to imagine the tenantry's life before the Clearances concluding: 'You would be among the poorest peasants in Europe but you were happy with your lot.' At the MoI, it is the voyage on board emigrant ships that the visitor is invited to experience – 'Imagine lying in a space not much bigger than a double bed. You are not on your own. You share this space, a glorified bunk, with your whole family …' – the experience is also kinetic as, the visitor is led through a passage meant to represent the cramped and oppressive interior of a ship, a narrow corridor physically signalling the transition from the display related to the Clearances period to the emigration section (Figure 5.4 below).

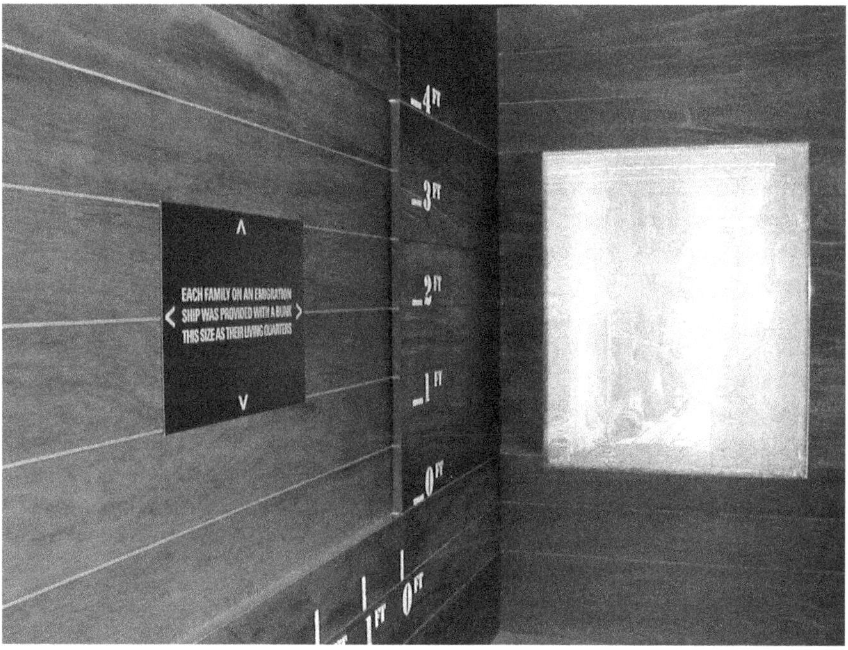

Figure 5.4 Museum of the Isles, Armadale, Skye
Source: Laurence Gouriévidis 2007.

Empathy and identification are forceful rhetorical tools often relying on sensory aspects which cannot fail to induce a strong emotional response in the viewer, not least when dramatic scenes are reconstructed with tableaux and sound effects, as was the case in Timespan's and *Aros*' former exhibitions which were based on elaborate stage-management. These scenographic and rhetorical devices help frame the visitors' reception of the past, serving to graphically contextualise contemporary sources or dramatise narratives. At Gunn HC and Ullapool M&VC, to name but two examples, poignancy and affect are communicated through the use of poetry and literature translating the emotional response of evictees. Irrespective of the anachronistic inclusion of N.M. Gunn's *Butcher's Broom*, and perhaps more tellingly, what the use of these sources vividly demonstrates, is the power of artistic creations in conveying the human dimension of a social process and ultimately its imprint on imagination and collective remembering – the force of post-memory.[4]

Narratives, partly owing to the legacy of the Clearances with their emblematic use signifying trauma and injustice, are also modulated by other crucial elements often found in museum and more widely public memory discourse: the use of the notion of victimhood (Lowenthal 1996: 74-78).[5] Conveyed through the evocation of scenes of eviction or other oppressive proprietorial tactics or through the ordeal of an emigrant's voyage and hardships of a settler's life, it might give the overall narrative an inflection tending towards loss and distress and insist on a heritage of subsequent affliction. Alternatively, it might underscore survival, resilience and determination and insist on progress and achievement. Disparities in terms of focus and narrative strategies used to impart these messages often bring to light underlying purposes and forces. For instance, in districts recounting the events of the Land Wars, choices are made to either foreground or downplay frictions and the ensuing Crofters' Act. They are strong features of exhibitions at Colbost FM, Ullapool M&VC, Kilmuir's SMoIL and formerly *Aros* with, at times, triumphant hues. At Kilmuir, the Crofters' Act is described as 'the first major step to curb landlord's power' removing insecurity and encouraging crofters to undertake improvements. Similarly, at *Aros*, clearances were presented as the culmination of a long history of injustice and the survival of crofting attributed to the actions of crofters in the past such as the Battle of the Braes. By contrast, in another Skye museum, the MoI, the Land War is distinctly upstaged by the emigration resulting from clearances and other factors, and nothing is revealed about how the troubles affected the MacDonald estate – where the Battle of the Braes was located – or other parts of the 'Isles'.

Interestingly Clan Museums, namely Gunn HC and the MoI, display telling nuances in their use of victimhood. Central to their overarching narrative of dispersal foregrounding the contribution of the clan around the word, are the notions of cohesion and kinship. With Gunn HC, cohesiveness is not affected by tales of evictions and the perspective adopted is that of a victimised group 'who

4 See Chapter 1.

5 For discussions of the use of victimhood in memory discourse see: Todorov (1995: 55-7), Ricoeur (2000: 97-105, 1998b).

suffered more than most' at the hand of uncompromising landlords, in particular the Sutherland family. The narrative reads like a lament, deploring the fate of a subjugated group broken and betrayed. Conversely, in the MoI, the same approach sits uneasily with the evocation of clearances, such as those, widely publicised, which were enforced during the mid-century famine period on Lord Macdonald's estate, and the narrative is deflected away from Skye and the properties of Clan leaders to encompass the whole of the Highlands and Islands. Before 2002, this was achieved through a schematic chronological table tending to obliterate causal relationship and details and to heighten a sense of inevitability. Since then, although the visitor is told that 'in many cases, clearance was on the orders of the landlord chief', the summon of eviction illustrating the process relates to a property belonging to the Baillies of Dunain from the parish of Inverness but including many Macdonalds among the evictees.

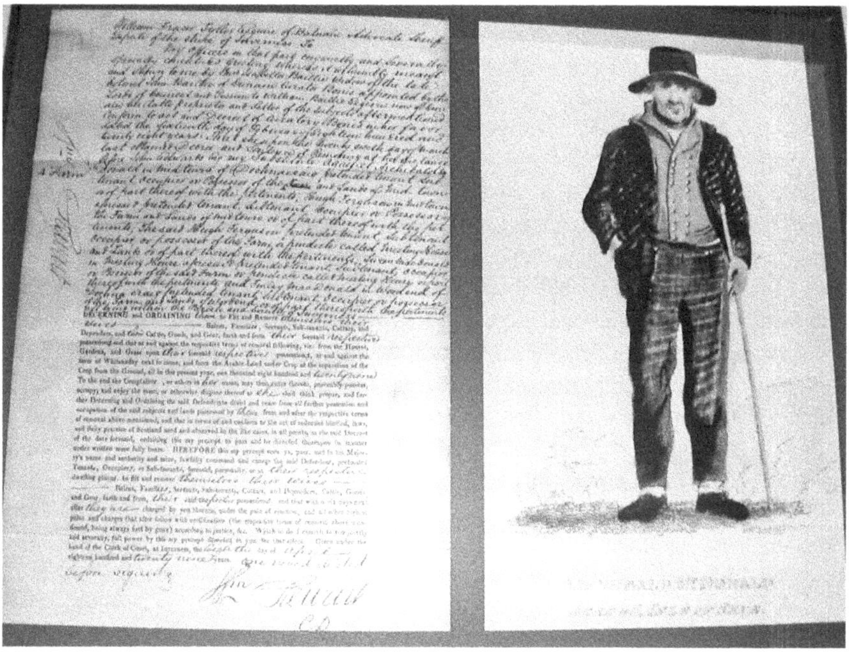

Figure 5.5 Museum of the Isles, Armadale, Skye
Source: Laurence Gouriévidis 2007.

The widely censured famine evictions at Sollas, Suishnish and Boreraig, are simply forgotten giving pride of place to the torment of the voyage. In Clan museums, as the MoI demonstrates strikingly, glossing as it does over clan disunity and fractures, the sense of the local is superseded by an appeal to the 'global', fostering a tribal clan identity and serving a world-wide community. In like manner, other venues, such as Arran HM, Glengarry HC and, most glaringly of all, the MoS, where emigration is also the leading strand, victimhood is clearly related to experiences in the home country; when the narrative shifts to host countries, it pales besides accounts of settlers' fortunes and triumph against adversity, but serves as counterpoint enhancing the tune of resourcefulness, resilience, not to mention freedom from feudal practices. Departing from museums in the emigration strand, the NMoRL, where Clearance is an overarching thematic trope encompassing Scotland as a whole and denoting migration or 'resettlement', victimhood is woven in through the destruction of a familiar way of life, landmarks and network of social and affective connections. Whatever their intended objectives and irrespective of the narrative devices used, many of the museums surveyed show a distinct penchant for the use of imagination and affect in conveying a sense of the past to visitors; this may not necessarily limit historical understanding nor hinder questioning, provided, however, that layering of information and variety of perspective are also integrated.[6]

Contested history and Clearance narratives in museums

Museums have often been accused of conveying a simplified and drastically compacted version of the past (Schlereth 2004: 340-41) and in response to the view that the interpretations they constructed were seldom 'presented as precisely that – one among several possible interpretations' (Uzzell 1998: 15), many have redefined their strategies and experimented with new approaches. In the case of interpretations of the Clearances, the aim of this study is to examine the extent to which the contested nature of the Clearances had been a concern for museum volunteers and curators, and filtered into museum narratives. Some of the questions formulated in the survey gave thought to the fact that museum exhibitions may be both reflexive[7] – conveying a sense of the difficulty of interpretation – and reflective – indicative of present pressures, be they local or national and even broader; they stemmed from a vision of museums as institutions fundamentally embedded in specific social milieux and times.

One of the questions directly addressed the notion of the Clearances as contested: 'there is not always widespread agreement on the interpretation of the period. Is it portrayed as a contested issue in the display?' Whilst being broadly associated with the cultural and political legacy of the process, the question was also conditioned by the idea of museum exhibitions as reflexive, in the hope of unveiling some of the dilemmas faced by the interpretation of such a highly charged event.

---

6 For a discussion of the use of imagination in heritage, see Gregory and Witcomb (2007).

7 For a discussion of the notion of reflexivity see Nederveen Pieterse (2005: 163-79).

Arran HM: Limited to material available and not really interpreted or commented upon.

Gunn HC: No.

Strathnaver Museum: We have tried not to do this but it is such an emotive issue locally that it is nearly impossible to avoid.

Timespan: Briefly, as local feeling is extremely strong on this subject.

Mallaig HC: Opposing opinions are presented but not highlighted.

Ullapool M&VC: It's not really dealt with as an issue. Facts are presented. The emotive stuff comes through the contemporary sources.

Seallam! VC: Not really as the point is taken that Clearances did happen at specific times.

Glengarry HC: No.

In keeping with traditional museum conventions, few of the respondents seemed to conceive of exhibitions as opening avenues for questions related to the (re)construction of the past – in the museum space and more generally. Seemingly, none linked the notion of contest with the production of history or the nature of history as a discipline. Yet, significantly, some mentioned the implications of interpreting the Clearances in the light of the strength of local opinion, attesting to the role of local museums in civil society, presenting narratives of the past textured by current concerns and certainly consonant with the views of those involved in elaborating the exhibition and, as argued by some, the wider local community.

The reactions to the question addressing the notion of 'objectivity' reflected similar internalised conventions of museum practice, whilst also casting light on the ways in which collective imaginaries impinged on interpretation. To the question: 'how important is the question of 'objectivity' in such a presentation? Can, in fact, museums/centres adopt an ideological position?', a few respondents provided unambiguous statements:

Strathnaver Museum: I think that it is very important to be objective. Viewpoints and knowledge are constantly changing and improving. A museum should be able to rise above current/personal viewpoints which can distort the facts.

Gunn HC: important.

Seallam! VC: very important but a museum should present facts not ideologies!

In the majority of cases, whether or not the exhibitions surveyed did present the Clearances as part of a continuum whose effects affected present-day Highlands and Islands, informants brought out the force of the legacy of the period in collective memory, both locally and more broadly, implicitly conveying the conundrum of interpretation. Furthermore, the agency of visitors was, as a result, acknowledged either in terms of the 'preconceptions' that framed their knowledge of the past or their ability to process the information distilled in museum spaces according to their own context and position.

> *Arran HM*: we leave the viewer to make up their own minds on the material – though we don't perhaps give sufficient coverage to the landowner's perspective.
>
> *Timespan*: very difficult in small communities with strong emotional involvement.
>
> *Mallaig HC*: many do. Often, however, this is unconscious and a reflection of the popular view of the Clearances.
>
> *Ullapool M&VC*: very important but hard to achieve because images and facts have already been used emotionally so much that no visitors are ever starting without preconceptions. It's even harder given strong feelings of trustees in a locally owned museum.
>
> *Glengarry HC*: preferable to offer facts so far as are known, and allow viewers to form their own judgement. If an ideological position is adopted, then the existence of diverging views should be acknowledged.

Nowhere is the 'people's voice' in modern times less equivocally authored than in the smaller 'idiosyncratic' museums, such as Colbost FM or SMoIL, both situated on Skye, where information is packaged eschewing institutional formalism. There, text panels are not designer-led and the narrative bears the individual imprint and authorship of the museums' owners with links between past and present strongly drawn in terms of cultural policy and land occupation – if, of course, time is taken to read the information on display. Both museums give interesting insights into their owners' visions of past issues and the way they affected the island over the twentieth century. SMoIL displays Jonathan MacDonald's hope in the outcome of evolving crofting legislation and in the development and promotion of Gaelic cultural and linguistic policies, whilst Colbost FM, is unique in relying almost exclusively on a series of press cuttings addressing controversies which have captured Peter MacAskill's interest: the highly contentious sales of the Cuillins, or of Waternish. His words are worth quoting as they neatly encapsulate what many venues never state so explicitly:

> The right and wrongs, claims and counter-claims of landlord and crofter seem to be destined to go on into unforeseeable future. But the bitterness which is passed on to each succeeding generation does not do the Highlands and Islands any good. It is hoped that the day will come when landlord and crofter relations are improved to the point that all will be striving together for the general good of the area.

In both venues, what comes to the fore is the commitment of owners whose positions are meant to incite the visitors' curiosity and critically engage with them. Both are early museums and yet they point towards an important civil role which some modern venues with wider constituencies now attempt to fulfil more formally and museological practice to endorse: the museum as a public forum for debate and controversy, as a place where the visitors' own response to past events and processes is prompted. As stated earlier, museums have come to be conceived of not so much as places sanctifying a version of the past and presenting a master narrative, but as loci where disputed issues can be raised and competing claims aired, provoking reactions and questioning. This was not a social function widely expected in early museums whose main focus was the record, collection and conservation of past objects.

Artefacts and the staging of Clearance narratives

Museum narratives are strongly inflected by arrangement and control of voices, rhetorical devices such as empathy and imagination, or use of tropes; in some cases, they also comprise artefacts (see Table P.2). The extent to which these enrich understanding of the period is not only premised on the type of objects used, but also on the way in which they are woven into displays and texts. In her analysis of the link between museum objects and the acquisition of knowledge, Jordanova delineates the processes at work in visitors apprehending artefacts:

> In order to gain knowledge from museums, viewers, whether they are aware of it or not, *both* reify the objects they examine, treating them as decontextualised commodities, *and* identify with them, allowing them to generate memories, associations, fantasies (Jordanova 1989: 25).

Further on, she also emphasises the limitations of the visual in conveying aspects of the past not so easily grasped through senses such as, for instance, a legal system (ibid. 26). As much as analyses of the organisation and display of collections of artefacts have cast light on the classificatory and categorising power of museums and, in this respect, their conventionalised practices have often been critiqued, they have also demonstrated the extent to which museums also rely on the sense of wonder and magic which exhibits are frequently meant to elicit, awakening imaginative responses (Joly 1998: 72-79). Museum practice regarding presentation and interpretation of artefacts is very much a matter of objectives and methodology, both conditioned by the identity a museum wishes to project and develop as a curating institution. For instance, in some museums, artefacts are carefully arrayed for effect, and conspicuous labelling is seen as interfering with the contemplation and appreciation of objects, which might have been selected because of their aesthetic value, their craftsmanship or their symbolic power. Equally, doubts have arisen about the pedagogic or basic instructive quality of artefacts, largely because this stance assumes that the meaning they hold is universal and unique, that the reactions and thoughts they prompt are common amongst visitors, irrespective of their individual experiences and identity – whether measured in terms of gender, class, ethnicity, age etc. It is a stance which ignores both the socially and culturally determined nature of knowledge and perception along with the polysemic character of objects. Whether museums use objects as illustrations of narratives or whether a narrative is produced through the combination and arrangement of artefacts, museums are in the business of assigning meaning, and consequently, it has become more common for modern museology to strive to exploit the polysemic nature of artefacts and open the routes of understanding through inventive presentation and mediation (Hooper-Greenhill 2000: 76-77, Sherman and Rogoff 1994: ix-xx, Crane 2006).

Folk museums have often been used as prime examples of venues where understanding of the past is compromised because it is conveyed through the prism of nostalgia and romance, and is limited to the functional dimension of artefacts. In them, visitors 'are most likely to learn about the workings of the plough than those of the laws that governed life in the country (Lumley 2005: 24)'.[8] However, it is often in folk museums that the Clearances are presented or, alternatively, folk-related relics – be they exhibited in more or less elaborate period rooms or in display cases – complement or are found alongside narratives of the period. Does this corroborate the belief verbalized by the Scottish ethnologist, Alexander Fenton?

> [Material culture] has a real and fundamental contribution to make in local history studies, in indexing in the round the way of life and economy of an area, and the course of change and development over as long periods as documentary and archaeological sources of information will allow (Fenton 1985: 13).

It is certainly an ideal that I.F. Grant shared, when she was building her collection for the HFM, but one which she knew folk museum exhibitions could not necessarily fully achieve for the general public, as her reference to the changing components used for the making of local tools illustrates; they could only translate a particle of the full gamut of nineteenth-century socio-economic transformation:

8 See also Kavanagh (1991), Bell (1996).

> A folk museum can only try to indicate the vast changes in the ways of life of the country people that the Agricultural Revolution brought about – from a condition of extreme poverty, utterly dependent upon the cruel variation of an unpredictable climate, yet with standards of behaviour that we look back on with envious admiration. In collecting their implements of husbandry we can show change from locally made wooden implements with the minimum of iron reinforcements to imported implements, mainly of iron and steel (Grant 2007: 49).

She also underscored the necessity of supplying visitors with 'as much background information as one possibly can by pictures, maps and written matter of anecdotes, bills of sales, description of the methods of use and anything else that one can think of (ibid. 176)'. However, in the majority of folk displays surveyed, material culture is seldom synonymous with socio-economic or cultural changes. This point is borne out by the recurrent separation of Clearance narratives from folk artefacts shown in separate zones between which visitors journey, and by the reactions of the SMoIL respondent to the section of the questionnaire dealing with the interpretation of the Clearances period, which was left blank, suggesting that the historical context is subordinate to the sensorial experience of the past. Yet, at the SMoIL, the narrative of the period is one of the most extensive; in the 1990s a pamphlet version was available for sale at the museum and, more recently, for a while, it was made accessible to virtual visitors via the internet (J. MacDonald n.d.).

Among the exhibition surveyed, folk material culture, depending on display arrangements and surrounding information, betoken different interpretive agendas and discursive fields. In Highland and Island museums, it is illustrative of regionalism or localism, emphasising the cultural and environmental specificity of place through building techniques, material and layout, through tools and domestic utensils. Conversely, at the MoS, little prominence is accorded to regional particularities in the 'Living on the Land' section and, as mentioned in the previous chapter, artefacts are not explicitly synonymous with geographical disparity, but in line with a notional all-encompassing folk culture. Highland and Island material culture is subsumed within the representation of a broad national folk life and the display projects the rural culture of the nation rather than the culture of the different people constituting it. This may partly explain why the Clearances are not featured in this part of the exhibition. In the NMoRL, the interpretation of the Clearances, which also resorts to national vocabularies, is not articulated around artefacts, but around text, audio-visual means and a few pictures. However, it was never intended as

> … a "folk" museum. The folk will be there, not das Volk, nor Mrs Thatcher's one dimensional "ordinary people", but as part of a social picture that has various layers and changes in different places at different times (Sprott 2000: 112).

Whether the museum is successful in projecting regional 'layers' is open to question, but what matters is that, clearly, in the two nationals, folk material culture was deemed insufficient or inadequate to impart what, in essence, meant transformation and manifold disruption.

Mark Jones' commentary on the principles adopted by MoS' staff as regards perspective and temporality gives us valuable insight into the choices which guided the type and use of artefacts in support of its Clearance narrative:

> The broad chronological structure makes it easier to understand the whole. The long themes within it reflect the evidence of the objects themselves, that the material culture of everyday life, and by inference the *patterns of existence*, remain unchanged for ordinary folk through periods that historians have tended to regard as very different (Jones 2000: 9, my emphasis).

To some extent, this also applies to folk museums or the use of folk-related relics in displays, where material culture often acts as the juxtaposed evocation of a perennial way of life next to narratives foregrounding precisely the opposite, namely uprooting, dispersal and breaks in 'patterns of existence', be they related to sociability or occupational activities. Within this context, at the MoS, where artefacts were the bedrock of the exhibition, it seems that the representation of the Clearances offered challenges and was driven by the assumption that they had left little or no material traces: 'What about the Highland Clearances with virtually no surviving objects?' (Calder 2000: 43).[9] Yet, the use of Clearance memorabilia seems to bespeak curatorial control and narrative choice rather than lack of artefacts. Many of the items showcased in 'Scotland and the World' are re-contextualised within a meta-narrative of emigration despite being loosely related to the process. In contrast, they were emblematic of, admittedly iconic, events and people linked with the Clearances period such as the Sutherland family or Sheriff Ivory, but were also revealing of social relations and conflict, or economic circumstances, for example the artefacts rewarding the best specimen of kelp in Ross-shire in the 1820s or the arrest of a land agitator on Skye in the 1880s. In the best of cases, the display may provoke questions and act as a springboard for inquiry in visitors feeling short-changed and wishing to discover more of a story, the bulk of which still remains to be told.

Museum artefacts are, by essence, re-contextualised by virtue of their use within an exhibitionary space and also their association with historical figures, events or processes framed by a thematic narrative line. Clearance narratives certainly rely on intrusive interpretive means, and folk material culture in particular, hardly speaks of social relationships and estate policies, if left undisclosed through interpretation. There are several ways of reading the use of folk material culture in the museums surveyed and how it might be perceived by visitors. If it is segregated

9 Incidentally the theme of mobility also seemed to present difficulties in terms of material culture, see, in the same volume, Forsyth (2000: 117).

from Clearance narratives, then it is likely to be, at best, functionalist linked with the characteristics and purpose of implements and utensils and their everyday uses, at worst, simply essentialist, conjuring up a vague sense of past times and past people, – this depending on level and quality of information and arrangement. Severed from or lifted out of their functional context and cultural milieu, artefacts may even reinforce a sense of otherness and discourage connections with history as a fluid and living process. If featured in close proximity of Clearances narratives, then folk material culture may perhaps operate at a metonymic level and invite visions of rupture or, at least, relocation; it may also elicit questions as to whether such activities were disrupted by the Clearances or how much of these practices were retained or adapted. Additionally, the mnemonic value of artefacts, mentioned by Ludmilla Jordanova, is one which is frequently exploited and was undeniably significant when I.F. Grant first exhibited her collections. In her writings, she repeatedly recalls how a *cruisie* light unleashed a flood of fond memories in an old lady catching sight of it (Grant 1989: xii, 2007: 134-35), memories of people by then deceased and places lying in ruin. Yet as each generation passes and becomes increasingly urbanised, imagination and affective empathy cease to evoke this form of experiential recollection related to lifestyle, and it is doubtful whether today's audiences engage with such artefacts with the same sense of wistfulness for a lost past. Nonetheless material culture also has the potential to tell of past cultural mind-sets and preconceptions when, for instance, used in conjunction with primary sources emanating from outside observers. At the MoS, such social commentaries are found in the words penned by Thomas Pennant and tell of the ways in which the living conditions of the Scottish peasantry were perceived: 'The house of the common people in these parts are shocking to humanity; (…) they look at a distance like so many black mole-hills.' The contemporary social and cultural perception of objects and material life is, however, not an aspect to which Highland and Island folk museums draw their visitor's attention, ultimately limiting the significance of their message and collection.

Archaeological sites and the staging of Clearance narratives

Material evidence, most frequently related to folk culture, rarely seems successfully and seamlessly welded into Clearance narratives, partly because of the inherent immobility that displays tend to exude, expressing 'stability in the face of change' (S. Macdonald 2007: 276). To provide visitors with a visual signifier of removals and changes in land use, two museums guide them to trails round pre-clearance archaeological sites. From Strathnaver Museum, where archaeological artefacts recovered from surrounding excavated sites and models of houses illustrate the narrative, visitors are directed to the sites of Achanlochy and Rosal situated down the strath, and at Glengarry HC to the nearby ruins of Daingean. There, visitors can gain first hand experience of the scale of houses and settlements with the lie of the land and surrounding landscape, and appreciate past ways of life and changes through interpretation; the sense of place and the sense of the past are meant to commingle, becoming closely entwined to conjure up a vanished way of life. Yet, although all sites try to give a glimpse of what the lives of the local inhabitants were before clearances occurred, their differing approaches to interpretation bring out contrasts between interpretive styles and objectives with Strathnaver villages and Daingean showing different emphases. Additionally, as the interpretation of two Strathnaver sites, Rosal and Achanlochy, underwent revision at the turn of the century, variations in interpretive strategies and methods reveal changing practices and contexts.

> As heritage sites are managed, the performance of what is chosen to be remembered and forgotten about the past is enacted, and its conservation and presentation to the public will affect "sense of place" and other experiences (Smith 2006: 94).

Through the analysis of these three sites, changing priorities, evolving knowledge and the interpretation of heritage as a social act embracing a variety of priorities come to light. By means of preservation and subsequently interpretation, ruins are given cultural value and meaning and those often speak to local agendas as well as wider concerns.

In the three instances, the heritage groups involved in the museum venues were either directly associated with devising the interpretation or, in the case of Rosal, consulted by the company in charge of producing the panels, Touchstone Heritage Management Consultants. Both Daingean and Rosal are sites managed by the Forestry Commission. When, in 1961, the Commission was planning to plant the hill around Rosal, it was decided to exclude the site of the settlement from planting and the investigations led by Fairhurst laid the foundations for initial interpretation. Given the fact that, at the time, few such depopulated settlements had been scheduled and even fewer interpreted, the Commission's initiative seems remarkably novel for its days (Hingley 1993). By the 1990s, the Forestry Commission had produced a set of guidelines with respect to archaeological conservation (Forestry Commission 1995). Daingean was re-discovered much more recently, in 1999, during clear-felling operations and was saved from replanting. Hidden under thick forest cover, its exact location had been forgotten by local inhabitants and few of the younger generation knew of its existence. The recovery reached the columns of *The Scotsman* under the headline: '"Lost" village resurrected for future generations' (J. Ross 2005). The 'resurrection' and opening up of sites to the visiting public certainly alert visitors to features in the landscape and traces of the past that might otherwise elude them, but more often than not they inscribe memories of past events within narratives smoothed out and recast in the light of present priorities.

The most striking examples come from Strathnaver where the revised interpretations of the two sites coincided with the establishment of the Strathnaver Trail, a protracted project fully open in the spring of 2004; the 40km long circuit takes visitors through 16 sites of archaeological and historical interest, of which Achanlochy and Rosal are only two of the five pre-clearance settlements recovered. The trail's meta-narrative hinges on the area's long history of human occupation, dating from the Neolithic, whose various layers are uncovered along the stopping points. Prior to the trail, both Achanlochy and Rosal had been interpreted for the public and comparisons between the two phases of interpretation display interesting shifts of emphasis. At Achanlochy, the first version, produced in the early eighties, was limited to one short descriptive panel mentioning the year when the site was evicted, the number of families removed and the identity of the factor, Francis Suther. The remaining information described the type of houses inhabited. It was brief and strictly noncommittal.

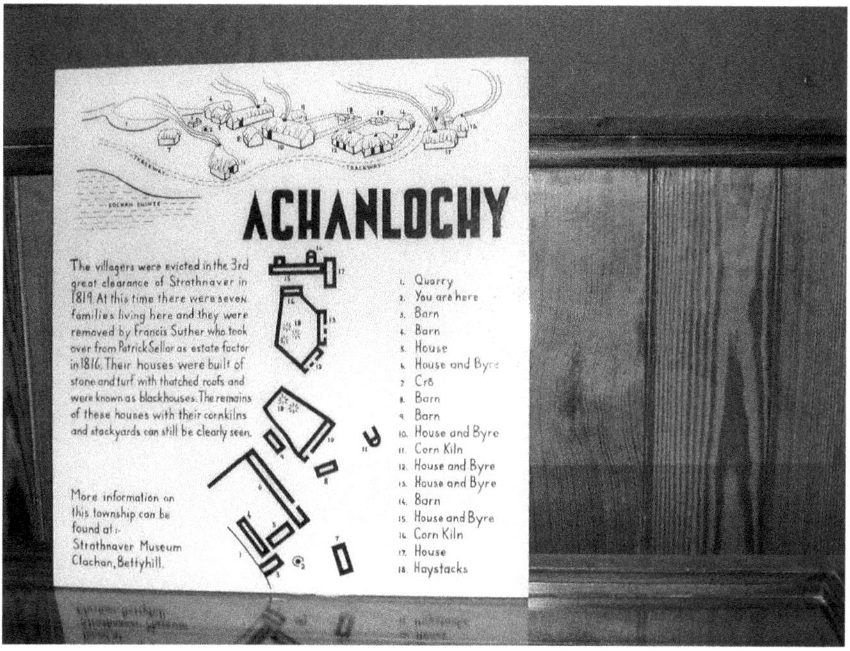

Figure 5.6 Achanlochy initial plaque, copy on display at Strathnaver Museum, Bettyhill, Caithness
Source: Laurence Gouriévidis 2007.

The second version is more extensive with two plaques, the first one offering a general presentation of Achanlochy, as typifying nineteenth-century Sutherland villages and the second focusing on its clearance.

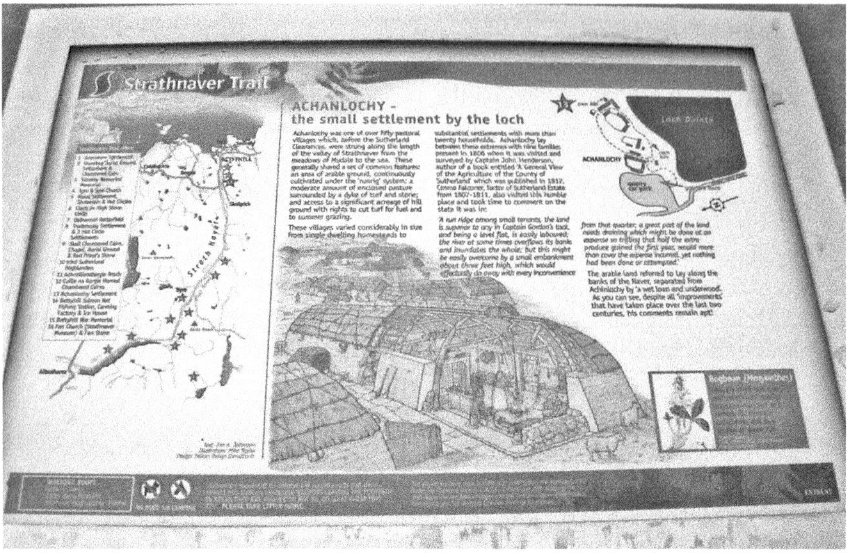

Figures 5.7 and 5.8 Achanlochy, Strathnaver Trail
Source: Laurence Gouriévidis 2007.

In contrast to the conciseness of the previous version, the recent one follows a conventional story-telling format with an intrusive narrator commenting on the personalities of the successive factors, on the methods used by Suther to carry out the eviction and on the reactions this provoked, before concluding with the doleful words: 'Achanlochy, as a community was gone forever and all that remains is what you see around you – the footings of houses, stackyards, barns and cornkilns decaying in to the hill side.' The remains of the village are framed by the representation of its eviction 'brought back to life' through story telling, its history frozen in that final moment. It is a narrative driven by the actions and emotions of those ultimate events, rather than the archaeological quality and specificity of the site.

Whereas, for Achanlochy, members of the museum heritage group were responsible for both interpretations, for Rosal, the Forestry Commission took charge of the interpretation with consultation with museum members and many other individuals. Yet, the two sites display similar narrative shifts towards more conscious use of story-telling devices, Rosal being far more visitor-oriented and striving to appeal to a wider age-range through the personalisation of the history of the village. The trail round Rosal was first constructed by the Forestry Commission in 1979 and clearly owed much to Fairhurst's findings and approach which had been presented in article form in the late 1960s (Fairhurst 1967-1968). Borrowing from his themes and conclusions, it focused on the types of dwellings found on site, demographic characteristics, settlement layout or type of agriculture practised – although on this point it misinformed the visitor alluding to lazy-bed cultivation. It also echoed the questions he raised as regards problems of land hunger and the lack of evidence of burnings in the township – thus throwing doubt on the story of Patrick Sellar's brutality. In terms of tone and assessment, it underscored the predicaments of living in such a remote location and harsh climate. Recurrent allusions to famine and times of scarcity, to a sparse and unvaried diet dampened any illusion of rural bliss, as these words did with respect to land use:

> One point which is not at all clear is why the inhabitants of Rosal did not make more energetic attempts to increase the area of arable land on which they were so dependant. It may be that with their primitive tools and means of transportation and the considerable distances over which they had to carry fertiliser, they had reached the limits of what they could maintain.

If 'scientific' distance was the keynote of the interpretation, it also imparted some sense of the process of historical and archaeological research including critical stance and interrogation of evidence. In contrast, the revised interpretation jettisons the professional voice of the expert and relies on empathy, imagination and emotional involvement by constructing a tale in the voice of several imaginary tenants, among whom a recently widowed mother. Their words, intermixed with some Gaelic, tell of the everyday life and daily activities of the inhabitants of the township, faced with the prospect of eviction, along the course of many months.

To convey an impression of the changes discussed, the following table presents the titles used in the two interpretive phases; in the most recent version, the use of pronominal forms and figurative language with alliterations, sound and word patterns, strive to 'humanise' the past.

Table 5.1 Rosal – interpretive changes

Initial interpretation	**Revised interpretation** **(Titles in both English and Gaelic)**
• Introductory plaque. The clearance of Rosal township • The Long House • Continuity of settlement • Population • Lazy-beds • Shielings • The peat bog • Food storage • Food supply • The township boundary	• Introductory plaque: trail and interpretive method • Rosal Clearance Township • The laughter of the children • Built by our own hands • We belong to this land • With the plough and patience • Back from the shieling • Sharing the soil, sharing the toil • Peat, wool and water of life • Spinning yarn and spinning yarns • From the pit to the pot • We are weary of the sound of southern men • Where are the people of Rosal?

Different though the perspective is, a closer look at interpretive texts dealing with similar themes makes it possible to compare like for like:

(Initial version): Food Supply

> The stony soil of the strath had to be cleared of boulders before it could be worked by the primitive *cas chrom* or foot plough which was the main cultivating tool. These boulders were collected into clearance mounds which are an obvious feature of this and many other primitive settlements. The community's existence depended on what could be grown and stored for the winter and the bulk came from the arable land. There was no surplus and any crop failures resulted in shortage or famine, several of which occurred during the nineteenth century. Cereal formed the staple diet until the introduction of the potato which gave relatively plenty but dependence on it was dangerous since heavy losses could occur through blight (…)

(Revised version): With the plough and patience

> Before we can plough, we have to clear the land of boulders. Then we build up the soil in rigs with furrows in between. *Cha bhi toradh gun saothair* – as we say there will be no produce without labour. The land here is quite fertile and we have no need for *feannagan*, what are called lazy-beds, where all that is lazy is the ground!

> When our faithful *cas-chrom* is put to work, it turns up so many stones that the piles grow bigger every year. Donald MacKay Bain says *clachan*, stones are our best crop. The *cas-chrom* looks so simple but it is clever in getting round big stones and roots of trees.
>
> It takes many days before the ground is ready for sowing and planting and *sinne na mnathan*, we women, work as hard as the men. We can plant *buntàta*, potatoes, which we rely on so much, sow the *eòrna,* barley or sow *corc,* oats for *brochan*, porridge. We try to use sloping land to help it drain.

As Figure 5.9 illustrates, in their updated version, each of the panels combine a main story-telling area and a smaller section with additional 'historical' information. For instance, in this case, the 'faithful' *cas-chrom* 'or foot plough' was described as 'a basic implement that needed great effort when used to break up poor soil.'

Figure 5.9 Rosal Trail
Source: Laurence Gouriévidis 2007.

The two versions presented could not be more different in tone, message and, in some measure, target audience. In their treatment of the post-clearance era, whereas the first one emphasised developments at home with the 1886 legislation and later establishment of the Crofters Commission, the latter foregrounds Rosal inhabitants' move to other areas, zeroing in on emigrants and ignores the land legislation introduced in Scotland. Interestingly, the final panel brings visitors back to 'reality', abandoning the characters whose daily lives they had so far shared and whose voices had guided their journey into the past. It is the 'reality' of migration, and especially the emigration and life of the Highland diaspora abroad. In keeping with the stance found in museums where the emigration strand is found, the celebration of a pioneering and entrepreneurial spirit dominates:

> People from all over the Highlands of Scotland contributed greatly to the growth and success of North America and southern hemisphere countries such as New Zealand, in agriculture, industry, commerce, education and politics. The spirit that kept the fires of Rosal burning for many centuries lives on across the world.

Arguably, interpretive changes may reflect efforts at making the trail more visitor-friendly and inclusive in terms of class, gender, age and geographical origins. If the story incorporates the voice of the people, with particular regard to women, and in the process delineates the region's cultural and linguistic specificity, it does not strive to achieve multi-vocality as regards contemporary standpoints. Variation in viewpoint is strictly confined to the visual split and narrative zoning between fiction and reality, between story-telling and the provision of brief factual details of a historical or archaeological nature, little else percolates. The approach adopted – the personalisation of past lives – is a technique shared by many heritage centres, relying on conveying a sense of the past through vicarious association, not unlike the effect sought by many writers of fiction portraying the human drama behind history. However, if the clearance of Rosal is anticipated and the subsequent process of migration is featured, no scene of eviction is enacted for the sake of dramatisation. Although earlier drafts of the project had included Donald Macleod's graphic account of brutal removals (Touchstone Heritage Management Consultants 1997: 19),[10] the trail's final version simply conveys the anxiety of the times and the stoic acceptance of what is to come.

'If it is at all possible, mention something gory' were the words describing the perceived over-emphasis of the initial interpretation on times of starvation and the practice of bleeding cattle and mixing blood with oatmeal. It was said to be depicting a version of the past which 'impose[d] all the common wisdom of the Highland Clearances on this one village', namely 'the histories of the improvers and their historians' (Mackay 1993: 46, 48). In the latest version, the practice of bleeding animals is presented as a means of 'mak[ing] a tasty blood pudding'. 'Gory' details, whether linked with social conditions or forthcoming events, do lurk below the surface of a narrative which, however, is intent on stressing a difficult yet immensely satisfying way of life adapted to its surroundings, a picture which is encapsulated in the recurrent image of the laughter of children. It is a way of life shaped by its environment, based on communality and attachment to a land occupied over many generations, but also responsive to socio-economic

10 At that point 'sound stores' were also considered to supplement some of the panels.

pressures as the allusions to cattle trade and seasonal migration indicate. If 'the destruction of old villages and settlements can be regarded as a condemnation of a whole way of life and a removal of the roots of the present' (ibid. 51), then the present interpretive trend underscoring past resilience, adaptability and eventually socio-economic viability, invalidates Improvement discourse along with property rights and reclaims a history celebrating the values and benefits of a collective order.[11] Still, because of its fictionalised form and its absence of any reference to the present of the local area against that of Rosal's far-flung expatriates, it is debatable whether the story manages to call the uninformed visitor's attention to questions relevant to the modern condition of the region and its inhabitants.

Whilst Rosal's interpretation is an attempt at creating 'living history' via a narrative which engages with the visitors' emotions and entraps their perception of the surrounding remains within a reading akin to folklore, at Daingean, the approach used was described as one striving to limit obtrusiveness and intrusion[12] in visual but also, I would add, emotional terms. This is notwithstanding the fact that the very act of interpreting is, by essence, intrusive as it determines the manner in which remains are to be perceived, by assigning selected meanings to them, and to a large extent, '"closes" [them] to alternative interpretation' (Basu 2000: 234) – albeit depending on the visitor's prior knowledge and background. At Daingean, the decision was made to interpret the site through a single introductory panel and a leaflet (Figure 5.10).

The panel tells of the history of the settlement 'occupied by ordinary families struggling to survive on the land whilst being caught in major historical events over which they had no control': Culloden and clearings. Demographic data provide the fabric of a representation of place associated with human occupation. Unlike Rosal and Achanlochy, although Daingean was cleared in 1785, it remained occupied 'on a reduced scale' by a handful of families of shepherds and gamekeepers until the inter-war period. The panel also draws attention to specific archaeological features found along the trail on which the leaflet expands: examples of rig and furrow agriculture, sheepfolds, various dwellings attesting to evolving building methods and the occupation of the area through to the early twentieth century, a corn kiln and enclosures. Personalisation and dramatisation were clearly not favoured strategies of communication and the interpretation does not attempt to take the visitor within the intimacy of a community, but uses the traces in the landscape, which the visitors are strongly invited to observe, as vehicles for ethnological information; the evidence thus discovered is meant to help them 'build up a picture of the way people lived in a township of this kind before the Clearances'.[13]

---

11 For a similar analysis of the Strathnaver Trail as a whole, see A.F.D. Mackenzie (2004).

12 *Project Design*, n.d. I am grateful to Hector M. Rogers for making this document available to me.

13 *The Daingean Trail*, n.d.

Figure 5.10 Daingean Settlement, Glengarry, introductory panel
Source: Laurence Gouriévidis 2007.

In all sites, the choice was made to adopt the broad notion of 'folklife' as a framework for the interpretation of material evidence and of the contours of human imprints left on the landscape. However, differing interpretive strategies result in the Clearances becoming emplotted in narratives of pre-clearance life in different ways, which, in some measure, can be linked to the dominant strand featuring in neighbouring museum exhibitions. The main focus of the exhibition at Strathnaver Museum is the process of clearance of many local villages by the Sutherland family. In the same vein, Achanlochy becomes defined by the very act of eviction which tolled its death knell whilst, at Rosal, the impending clearance of the settlement looms large over the portrayal of life's cycle and rhythm through seasonal activities; it is a way of life about to face destruction as suggested by folk voices abruptly falling silent in the last panel of the trail. These are 'stories' in which the aching memories of the past ooze into the sites. Conversely, Glengarry Heritage Centre is partly oriented towards the process of emigration from the region, of which clearance is one cause, and Daingean, where, in any case, life was not completely eradicated by nineteenth-century evictions, is used as a signifier of practices, techniques and methods used by people before they deserted the settlement. Clearance is thematically subordinate and does not bring closure to the narrative. Interestingly, in the plans outlining the Daingean project, the 1999 Burra Charter for the conservation and management of places of cultural significance was adduced to justify its conservation and maintenance.[14] The Charter, is part of ICOMOS' (International Council on Monuments and Sites) apparatus defining 'what heritage is, how and why it is significant, and how it should be managed and used' (L. Smith 2006: 87). Born in Australia, the Burra Charter pays homage to the notions of plurality and inclusiveness with respect to the involvement of 'people for whom the place has special associations and meanings' in conservation, interpretation and management (Australia ICOMOS 1999: article 12). Central to the Charter is the concept of 'cultural significance' which encompasses 'aesthetic, historic, scientific, social or spiritual value for the past, present or future generations' (ibid. article 1.2). It is premised on the social and mutable significance of heritage. In the context of the interpretation of clearances sites and more broadly Clearance heritage in its different forms, the Charter is an apt commentary on the changing nature of 'cultural significance' bound up with the groups to whom a place, artefacts, a version of history are of special concern at given points in time and the varied interests which might be directly or indirectly implicated in the story and its resonance. As this and the previous chapter have shown, the accepted view of the Clearances found in heritage practice in twenty-first-century Scotland is one in which the discourse of Improvement pales into insignificance.

---

14 *Project Design*, n.d.

Chapter 6
Museological Developments and Narratives of the Clearances

The Museums Association adopted the following definition of a museum in 1984. A museum is: "an institution which collects, documents, preserves, exhibits and interprets material evidence and associated information for the public benefit." Now, in many a heritage centre, finding "material evidence" – a "real object" – is like a game of hunt the thimble. But does this necessarily lessen the experience? And what is "material evidence"? (Heighton 1987: 61).

Heritage venues where narratives of the Clearances are found come in many guises, some built around material culture, some with few artefacts and much technological input, some in purpose-built edifices, others in listed buildings (see Table P.2 for a summary of the main characteristics of Clearances exhibitions). Their orientation is as much a matter of local choice as an indication of much wider trends affecting the museum world, which over time re-defined its methods and objectives. The purpose of this chapter is twofold: to chart museological changes and developments through the rich landscape of Highland and Island museums surveyed and examine how these evolving priorities impacted on Clearance narratives and their effects.

Early folk life exhibitions in Scotland

Many of the museums opening vistas into the past of the Highlands and Islands may best be described as rural history museums or folk life museums. In scholarly studies delineating their development, which, in Britain, mainly occurred from the inter-war years, the conservation of Highland heritage features prominently through the pioneering work undertaken by Dr I.F. Grant who was one of the precursors in the British folk museum movement (Kavanagh 1990, 1991, Bell 1996). She relentlessly scoured the Highlands and Hebrides to build the collection for the first folk museum in Scotland, the Highland Folk Museum, which she established single-handedly in 1935. Initially located on Iona in a disused United Free Church building, it was named *Am Fasgadh* – the shelter – because 'it was to shelter homely ancient Highland things from destruction' (Grant 1945: 3). As with most of her contemporaries in the movement, her source of inspiration was the Scandinavian folk museums which she had visited with her mother in Sweden and Norway, in particular Arthur Hazelius' two innovative and complementary initiatives founded in the late nineteenth century: Nordiska Museet, which presented formal collections and displays and focused on research and scholarship, and Skansen, an open-air museum, which exhibited buildings moved from various parts of Sweden and rebuilt on site and was an attempt to democratise the museum experience by appealing to a larger public (Cheape 1986: 114, Kavanagh 1991: 192). Hazelius and Grant were motivated by the same goal: to preserve, for future generations, the remaining material traces of ancient ways of living that were fast vanishing in the wake of the 'Industrial Revolution' (*Inverness Courier* 1930, Alexander 1979: 85). In museological terms, her approach, vision and inspirational models were not exceptional in the UK and mirrored a trend advocated by the Museums Association.[1] Interestingly, the folk museum movement was the epitome of a new exhibitionary spirit, a blend of new practices and themes meant to widen museum appeal by transcending class barriers, and promote, through recreation, good citizenship among the lay public, notably with the representation of a coherent national imaginary. In this programme, the success of and fascination for the many mock 'white villages' reconstructed at international exhibitions and fairs organised in the late nineteenth and early twentieth centuries, have led to analogies and comparisons between exhibitionary practices and ambitions (Coombes 2004). Amongst such villages were the various Highland *clachan* erected during the Glasgow exhibitions of 1911 and 1938.

The catalyst for *Am Fasgadh* was a temporary Highland Exhibition held in Inverness from 4 August to 20 September 1930 which I.F. Grant initiated and devised; she had hoped that this would spur on the founding of a permanent folk museum but it seems that in spite of the fact that the exhibition attracted visitors from afar, it failed to capture the imagination of the 'Invernessian' as the provost and Convener of the exhibition's Executive Committee, Alexander MacEwen (1930), deplored in the pages of the Inverness Courier in early September:

> Support [for the creation of a permanent museum] would depend ultimately on the interest taken in it by Inverness and the Highlands, and it would be a great encouragement if at this stage our own people would show that they are interested in the present effort.

This reported lack of local enthusiasm and interest in their Gaelic past seems to echo Grant's own sorry assessment that Scotland, let alone Britain, unlike other European nations, had failed to pay tribute to the past of their rural folks. This, she intended to reverse for the Highlands and the Highlanders, placing her endeavour within the wider nationalist discourse of the thirties in Europe.

> The movement is only one symptom, among many of the marked development of national and racial consciousness that has been such a feature of the last two decades, and it seems a strange thing, and rather a disgraceful one, that the

1 See the issues of the *Museums Journal* from its inception in 1901 to the inter-war period.

> people of Scotland, and above all, the Highlanders, with the tremendous pride of race that is so marked a characteristic of the Gael, and one of his greatest contributions to the spiritual make up of the Scots people, should have had neither part nor lot in this movement (Grant 1929).

Interestingly, Grant's words publicising the 1930 Exhibition help us to understand what her motivations were, how she conceived of her future museum and of Highland culture and heritage in the context of the inter-war period. They also point towards many of the functions that underpin the creation of museums.

She regarded the preservation of heritage as one means of defining and bolstering a sense of identity and certainly strove to inject a sense of confidence and pride in traditional Gaelic culture in the local public that she hoped her museum would attract. Her grammar recurrently betrays the deep reverence and admiration – tinged with a hint of romanticism – in which she held past Gaelic society singled out for the 'corporate life' and solidarity that had supported its people through hardship and were at variance with her own 'materialistic' and commercialised age (Grant 1930a: xxviii, xxxi). Her approach to conservation seems, as a result, more commemorative than solely intent on collecting and recording past artefacts and the museum she envisages is more akin to a sanctuary or memorial. This, she expresses candidly on the opening of the exhibition, alongside the Provost and Lord Lovat who also insist that the prospective museum must be a 'source of inspiration for the future' (*Inverness Courier* 1930). Herein transpire two complementary visions of the function of museums and the role of the past, that of a museum holding lessons for the future – a transmitter of values and beliefs – and, that of the museum as custodian of the past – safeguarding its remnants so that future generations might also see, experience and understand them. Her other guiding principle was to create a museum for all, specialist, layman and, most of all, the rural people themselves, to demystify the museum experience and, within this perspective, she saw artefacts not only as specimens documenting past practices and techniques, but as mnemonic tools generating personal memories, 'intimate remembrance of the long dead, beloved people whom they had seen using' (Grant 1945: 7). She also had in mind descendants of Gaels overseas in quest of their forebears' living conditions and, in the world of heritage, might be seen as a forerunner in her vision of Gaelic Scotland as a trans-national homeland. These goals were to be translated in exhibitionary and curatorial practices as her future museum was to combine the exhibition of classified artefacts alongside an open-air display of vernacular buildings.

Besides these objectives, what can also be perceived through I.F. Grant's conception of the 1930 exhibition are the negotiations underlying her interpretive work of the Highlands and the Highlanders in the early Thirties. Her exhibition was read, again by MacEwen, as conveying 'not only the military and *romantic* sides of Highland history but also the social and domestic life of the people in the past' (*Inverness Courier* 1930, my emphasis), a reading which betrays the mental universe which framed the construction of the image of the region and its inhabitants and, at the same time, underscores the other aim of the exhibition: to restore the world of the peasantry to its local socio-economic setting and, by so doing, reclaim its identity. It was no easy task, poised between the romantic pabulum inherited from Enlightenment urban intellectual imagining and nineteenth-century artistic mythologizing (Smout 1983, Womack 1989, Withers 1992), and their more recent version, no less demeaning, found in the Great Exhibitions where 'white villages' – Scottish and Irish – had become recurrent features offering visitors images of an edenised rurality in which 'realism was a low priority' (Greenhalgh 1988: 82-111).

It was the case of the 1911 Scottish Exhibition of National History, Art and Industry held in Glasgow, where one of the popular attractions was *An Clachan*, a Highland village which, in the same vein as its West African and Lapland counterparts found in close vicinity, showcased tartan-clad Gaels speaking their own tongue, performing their own arts and crafts in their native dwellings – a mixture of black-houses and 'but-an'-bens. Alongside the exhibition of primitive 'others', the Highlanders could be perceived as Scotland's own 'aborigines' (Kinchin and Kinchin 1988: 121), an internal exoticised 'other', displaying ancestral practices and culture and largely by-passed by the technological and industrial development embraced by southern Scotland – a vision which accorded well with the preconceptions held against the Celts in some quarters. However, the purpose of the exhibition was not simply to parade Scotland's industrial achievements, and it also had a strong national and historical dimension as its origin lay:

> … in the belief, shared by many that the time had fully arrived when Scottish history should be placed on a different plane from that which it had hitherto occupied in the education of the rising generations of Scottish children, as well as in the teaching of the subject in our higher schools and colleges (*Scottish Exhibition of National History, Art and Industry, Palace of History, Catalogue of Exhibits*, 1911: 3).

Its purpose was to 'raise a fund to the endowment of a Chair of Scottish History and Literature at the University of Glasgow' (*Scottish Exhibition of History, Art and Industry, Official Catalogue* (Industrial Section) 1911: 57). *An Clachan* itself was devised 'with the patriotic object of causing a greater interest in the Highland people, in their traditions and customs, in their beautiful Gaelic language, literature and music' (ibid. 69) and, even more importantly, to market Highland craft products and fare which could be purchased at the store and inn. In this context, the composition stood as the remarkable display of enduring, traditional craftsmanship and art transmitted through generations, yet this idyllic pastoral pastiche, listed in the same breath as amusements in the official catalogue, was also fee-paying which signalled to visitors that they were entering a mythical world of historical fantasy and had to suspend belief. The point was eloquently made by Neil Munro (1911) who wrote the introduction to the collection of articles on the social history of the Highlands, specially commissioned as a background to the village. If the vision of the Highlands and Highlanders at *An Clachan* pandered to long-ingrained stereotypes, either romantic or neo-colonial, the commercial motives of this *tableau vivant* in an age when tourism offered potential for development, are also undeniable – especially in view of the region's parlous economic condition. Such reconstructed vignettes also prefigure a line which, later, some 'living history' or 'folk park' exhibitions were to perfect and exploit through the formulation of interpretive and exhibitionary techniques.

As mentioned, although the 1930 Highland exhibition did not lead to the establishment of a permanent museum, I.F. Grant soldiered on and the little museum on Iona came into being. The remoteness of the island location and the growing exiguity of the building subsequently led her to look for an alternative site on the mainland. This she eventually found in Kingussie, and after facing many obstacles, in June 1944, the Highland Folk Museum was opened to the public. What set the Kingussie museum apart from other displays of rural collections was that it did not only provide a space where material culture could be collected, recorded and studied but, for the first time, it comprised of examples of vernacular architecture from various parts of the Highlands and the Hebrides where domestic artefacts were set in context. Before Kingussie came into being, her expertise and collections were given a fillip of acknowledgement during the Empire Exhibition of 1938, as she was responsible for furnishing and arranging a 'Highland homestead' in one of the Scottish Pavilions. In I.F. Grant's words, the display was reminiscent of the descriptions produced of such dwellings by late-eighteenth and early-nineteenth centuries visitors to the region (Grant 1938: 50) – whose vision, however, was refracted through contemporary philosophical or aesthetic canons. The 1938 exhibition also had its *An Clachan* with highland products for sale. The dichotomy between 'Highland Homestead' and '*An Clachan*', the former standing for 'the past history of *Scotland*' (my emphasis), the latter for 'romance' and 'nostalgia' (*Empire Exhibition Scotland* 1938: 121-22) was one laced with the same ambiguities as those manifest in 1911: the boundaries between history and myth, nation and region, tradition and quaint anachronism blurred and elusive. *An Clachan* was certainly devised around the collective imaginaries enveloping the Highlands and Islands as it was meant to 'raise many memories in the minds of returned exiles and will give to others some impression of the real old Scotland, the Land of the Gaels, the Scotland that is fast passing before the relentless onrush of modernity' (ibid. 121) – although a modernised cottage was also part of the stage-set. The charge of mythical or nostalgic representation of the past later levelled against folk displays is one which led to museological debates, as will shortly be seen.

If I have dwelt at length on the early approach to folk life displays, it is because many local museums have followed in I.F. Grant's footsteps, although in content, ethos and extent of collection, they may be different. Many smaller museums throughout the region have, since then, tried to imitate or borrow from this approach, either incorporating the interior of a room in otherwise multi-facetted exhibitions or reconstructing one or several vernacular dwellings. It is in some of them that Clearance narratives are found (see Table P.1). The owner of SMoIL on Skye takes pride in the fact that his museum was the first of its kind opened after Kingussie's Folk Museum (survey answer) and over the years the exhibition has expanded its collection of bygones housed in a growing number of thatched buildings. Both 'Clearance museums', Strathnaver Museum and Timespan, exhibit the domestic interior of croft houses. Interestingly, in the entirely refurbished version of Timespan, which opened in 2008, scenes illustrating the past are no longer the norm; they have been discarded when conjuring up events and processes such as local evictions and estate re-organisation, but have been retained to evoke material and social life through the reconstruction of 'croft street' with its shop, smiddy, croft byre and house. At the MoI the material evidence on display in the Clearances room 'Clansmen to crofters' relates to crofting bygones exemplifying domestic and working life. Addressing the national past, the MoS' 'living on the land' display is also directly inspired from HFM and the examples cited referring to building techniques and material are essentially drawn from the Highlands and Islands, yet the setting, as seen previously, is not used to reflect changing regional conditions, but rural continuity in Scotland at large. If the representation of folk life raises cultural issues in the modern-day Scottish context, it also led to wider questions, as the folk museum movement expanded in Britain and elsewhere.

Material culture and the interpretation of the Clearances

Whilst the growth in folk museums after the war marks a refreshing expansion of the social scope adopted by museums up till then, Tony Bennett, echoing Gramsci's misgivings, warns that:

> (...) the terms in which the ways of life of [subordinate social] classes are represented are often so mortgaged to the dominant culture that "the people" are encountered usually only in those massively idealised and deeply regressive forms which stalk the middle-class imagination (Bennett 1988: 64).[2]

Elsewhere it has been observed that early folklife curators, such as Iorwerth Peate in Wales or I.F. Grant, who 'complained at the ghettoisation of "folklife" were largely responsible for it'. Their collections of 'craft tools, ploughs and other "bygones"' begetting a 'dewy-eyed, romantic, reactionary contemplation of rural society' (Rattue 1996: 217). How then do the more recent heirs of the folk movement address the issue of the Clearances, an issue which can only invert this idealised image introducing notions of social and agricultural disruption and conjures up a rapidly changing rural scene? To what extent does material culture manage to convey the nature, pace and manifold effects of agrarian transformation of which the Clearances were emblematic?

2 This article is reprinted in Bennett (1995: 109-27).

Most of the museums surveyed illustrate their portrayal of local culture with artefacts relating to the daily activities of the 'people'. However, this section will mainly focus on the few which are dedicated folk life museums and, in fact, were amongst the earliest opened in the range (see Table P.1) – The Highland Folk Museum, Skye Museum of Island Life, Colbost Folk Museum and Isle of Arran Heritage Museum. Part and parcel of a UK-wide movement which saw their proliferation after the war (Harden 1956, Austin 1958), it might be argued that they mirrored the concerns which inspired I.F. Grant as they emerged against a backdrop of accelerating technological and social changes, their retrieval of the past offering a haven of stability and permanence in the face of frequently tense industrial climate and social unrest. Visitors, often from urban centres, would meander through collections of agricultural implements and cooking utensils, peer into box beds, smell the peat fire burning in blackhouses and indulge in the contemplation of a lost past.

The power of artefacts, of their arrangement and contextualisation has been amply demonstrated. Symbolically and emotionally charged, museum objects often act as visual shorthands, triggering recollections or images and thoughts acquired through education and social interaction. Likewise the systems of classification that direct their organisation may have a normative impact as in the case of evolutionary series tending to delineate universal laws of progress (Kavanagh 1989: 130, Bennett 1995: 147, Sherman and Rogoff 1994: ix-xx). The museums on the Isle of Arran, at Kilmuir and Kingussie present the most extensive collections of artefacts which are arranged around taxonomies, such as domestic life, agriculture, craft etc., often associated with specific buildings – croft house, barn, smithy, and weaver's shed at Kilmuir; cottage, smithy, stable, wash and milk houses on Arran. If, at Kingussie, owing to the extent of the collection, and to a much lesser degree at Arran HM, some sense of time-related change might be gained through a roughly diachronic presentation, at Kilmuir, specimens because of their diversity denote some pre-technological and pre-mechanised age and what mostly percolates is a timeless rural past seen through rose-tinted spectacle, the site re-creating an idyllic pastoral scene – prettified by such details as brightly-painted agricultural implements, adding to the artificiality of the picture (Figure 6.1).

Advertised as depicting 'the lifestyle of the crofting community of the island, a century or so ago', the visitor might be forgiven for imagining that crofters had a distinct penchant for ornaments and clutter, so packed with a multitude of china, musical instruments and other domestic or craft-related paraphernalia is the croft house. For the visitor, the owner's passion for collecting bygones might well translate into what seems like an accumulation of bric-a-brac and antiques (Figure 6.2).

Figure 6.1 Skye Museum of Island Life, Kilmuir
Source: Laurence Gouriévidis 2007.

Figure 6.2 Kitchen of the Old Croft House – Skye Museum of Island Life, Kilmuir
Source: © J. MacDonald.

With such exhibitions, the danger is, as ethnologists have recurrently pointed out, to turn museums into little more than 'a nostalgic peepshow into a largely fictitious past (Jenkins 1989: 123)'; pandering to the deep-seated sentimental streak in most visitors, folk life museum can easily fall into the trap of 'fakelore' or 'fakelife' (Fenton 1985: 33). The situation – not just at Kilmuir – is compounded by the fact that labels tend to be minimalist; the ubiquitous *cas-chrom*, peat spade, plunger churn and rotary quern disclose little about the particularities of a cultivation system, the social structure that underpinned it or the type of husbandry which prevailed and its characteristics and it would take a visitor well-versed in ethnology to grasp the scope of the socio-economic and cultural context of such displays.

By contrast, the no frills black-houses that are reconstructed, using traditional skills and techniques, at Colbost and at the HFM, in Kingussie and more spectacularly at its living history arm in Newtonmore – where a township based on the excavated settlement of Easter Raitts up the Spey valley has been erected – are certainly more successful in revealing the conditions of living and way of life of Highland and Island people.[3] For a start, they are more clearly dated, the house at Colbost is advertised as a nineteenth-century dwelling, whilst the township at Newtonmore is reported to date back to the early 1700s.

Figure 6.3 The Highland Folk Museum, Newtonmore, Strathspey
Source: Laurence Gouriévidis 2007.

3 See Atkinson and Holton (1973) for an assessment of the function and role of such museums in the British context in the early 1970s.

The roof structure, thatching material, shape and height of the buildings, the soot-stained ceiling and furnishings trimmed down to the essentials bring home the characteristics and constraints of the local environment along with the inhabitants' standard of living; likewise the tools and domestic utensils on site cannot be divorced from the surrounding terrain, climate and natural habitat. If, at a stretch, such replicas, particularly when visited on a damp cold summer day, may stimulate images of shortfalls of harvest and penury, it would still take a leap of the imagination for the visitor to come to grips with the full socio-economic reality of the community that they represent. More to the point with respect to the Clearances, buildings, artefacts and memorabilia – even when belonging to the Skye poetess of the land agitation – do not transmit the sense of flux and the reshaping of land, society and agrarian practices that accompanied the period. This points towards the limitation and methodological difficulties in using material culture with limited or fragmentary interpretation when attempting to tackle the events.

Similarly, it seems doubtful that archaeological artefacts dug out of cleared settlements, such as the broken vessels displayed in Strathnaver Museum, should have assumed a symbolic meaning without the contextual information provided, quite literally standing for the shattered community of Rosal where they were excavated. Hence the folk museums which do or did tackle the Clearances period have had to rely on more than the mobilisation of objects to tell the history of clearance, crofting or agrarian unrest. In SMoIL, in effect this leads to two separate storylines and exhibitions: one of pastoral tranquillity and relative comfort revolving on collected artefacts, the other, set apart in the aptly named 'ceilidh house'- where the experience of crofters is depicted and appraised in a conventionally-styled historical narrative 'history of crofting' – emphasising resilience and linking past events and present policies through text, pictures and other data. Likewise, a brief comparison of the changes which occurred over time in the exhibitions at the NMoRL at Kittochside, formerly Ingliston's Scottish Agricultural Museum, and at the Newtonmore site also brings out the predicament of using material culture to evoke historical processes and events. The move of the rural history collections from Ingliston to Kittochside was also a move in terms of approach, at least as regards the Clearances which only entered a narrative, built around the tensions between continuity and change, in the new venue. At Ingliston the slant was on ways of working the land in the different regions of Scotland and included the industrialisation of agriculture, particularly through the display of artefacts. At Kittochside, the social impact of changes is represented, notably through the Clearances but, as mentioned in previous chapters, at this point artefacts are jettisoned – apart from the emblematic *cas-chrom* – in favour of illustrated text panels and an audio-visual display. By contrast, at Newtonmore, where the crofting panels introducing the visitor to the concept and its history have now been abandoned, the Clearances do not seep out in an exhibition strictly restricted to 'living history', confirming the sense that, when it comes to social history, inclusive of structures and processes, material culture seemingly leaves many gaps and silences, even when brought to life with costumed interpreters.

Social history and multi-media exhibitions

The inadequacy of objects in engaging more than descriptively with social issues, in conveying more than a sense of painless change through material progress, in digging into causal links and introducing multiple viewpoints is only partly the reason why, in the majority of the exhibitions surveyed, artefacts were superseded by other means in particular texts and images – photographs, paintings or sketches – to project and interpret the story of the Clearances. Museum practice was also challenged in the wake of wider historiographical dynamics which rippled through the historical profession and beyond from the 1960s onwards. The advent and growth of social history with the publication of ground-breaking works often Marxist in inspiration, such as Thompson's *Making of the English Working Class*, the influence of the approach of the History Workshop and its focus on people's history, the recognition of the value of oral evidence certainly had an impact on the way museum curators and volunteers began conceiving of their method and means of reaching out to the visiting public. Already familiar with the notion of history-from-below, their focus was no longer to *describe* the way people lived but how their lives were affected by changes and how they responded to them. This meant moving away from an object-based approach in order to fill in the gaping holes that material culture left and using a combination of media that might best communicate the message to the visitor (Jenkinson 1989, M. O'Neill 1990, Coutts and Clark 1989). This changing intellectual climate – not limited to Britain – laying stress on concepts of cultural accessibility and democratisation acted as an empowering force at a local level encouraging local groups to reclaim their past through memorial activities such as the establishment of local museums, historical societies or commemorative activities.

In Gaynor Kavanagh's view (1990), the response of local museums to this social focus was from the perspective of 'community' history. The concept is presented as a 'useful means of identifying common interest and concerns: their manifestations, contradictions and changes'. At a practical level this also entailed enlisting the collaboration of local people (ibid. 66-67).[4] However, the notion of community is not unproblematic, as in the often-cited case of 'new' characterless commuters' towns, and rural museums projects could just as easily expose fissure as harmony within the communities involved in their planning, as has been shown for Argyll and Bute (Davis 1999: 31). Symbolically constructed, 'the reality of community lies in its members' perception of the vitality of its culture' (Cohen 1989: 118) and, when applied to the Highlands and Islands, it comes heavily

4 Springburn museum and Edinburgh's 'People Story' are good illustrations of the concept, see M.O'Neill (1987) and Clark (1988).

connoted since its use and definition in James Hunter's *The Making of the Crofting Community*.[5] Significantly, his research was the mainspring behind the 1990s exhibition set up at *Aros* on Skye and the exhibition's commentary transmitted to visitors through headsets as they walked through each display section was credited to him in the local press (*WHFP* 1993a).

In terms of inspirational drive, compounding the academic concerns which led to new interpretations of the Clearances, museum representations of the period using a people-oriented perspective also need to be associated with the literary production of the inter-war and post-war period and, above all, with the highly readable and compelling popular histories written by John Prebble (1969, first published 1963) and Ian Grimble (1962). Perpetuating a tradition started with Alexander Mackenzie (1986, first published 1883), their writings, by largely refurbishing previous contemporary accounts, not only made accessible a large number of sources which museums curators could tap but re-awakened a long-standing interpretive vein. In their works, the experience of evictees, crofters and emigrants came to life and could be reproduced within the confines of an exhibition. In one notable instance Ian Grimble was more than just a source of information as he played a crucial and active part in the founding of Strathnaver Museum; he 'conceiv[ed] the idea of the museum' in the 1950s, an idea which materialised with the help and support of a group of local residents when funding became finally available in the mid-1970s (Strathnaver Museum n.d. 10). In the museum's literature, Grimble's contribution to the study of the period is hailed as demonstrating 'passionate understanding of the viewpoint of the Gaels – something markedly lacking from the stance of establishment historians' (ibid. 9) and the museum's narrative of the events strongly articulates this position. Although the present interpretation was produced in the late 1980s, it bears the imprint of Ian Grimble's approach in *The Trial of Patrick Sellar*; they are both structured around discordant voices with a controlling narrator organising and assessing evidence (see previous chapter), ultimately judging and condemning the outcome of the policy, of which Sellar was one of the most salient symbols.

One of the purposes of social history was to express the stance of groups and individuals hitherto silenced and excluded – marginalised 'communities': women, ethnic minorities, the poor etc. In the context of post-colonial studies, what this resulted in was to give a voice to native inhabitants or 'first people' and reinterpret the history of the 'New World' not simply in the eye of the coloniser, as terra nullius giving pride of place to the endeavours and actions of explorers, settlers and administrators, but as conquered territories where expropriation and violation of ancestral practices and rights so often attended the colonising process; in the museum context, the perspective has generated much soul-searching, experimental work and controversies – particularly when meshed with issues of multi-culturalism and diversity (Cannizzo 1991, Simpson 1992, McIntyre and Wehner 2001). It is worth examining how the museums reviewed which foreground the emigration

5 See Chapter 8 for further analysis of the concept of 'community'.

strand, rose to the challenge. If, in a few cases, allusions are made to interactions with native inhabitants, overall little significance is placed on such a reading; expatriate Highlanders and Islanders are mainly viewed in terms of their immense contribution to the development and expansion of new overseas possession and to the prosperity of the British Empire or subsequent independent nations, conforming to traditional celebratory narratives of imperial conquest. More specifically both the MoI and the MoS zoom in on the fate of distinctive individuals or groups – trailblazers, ordinary settlers and successful traders or politicians – turning their personal stories into public memory; interestingly some of the examples selected tend to bring to the fore instances of 'friendly encounters' and even cultural hybridisation rather than the wider processes of appropriation and dispossession which territorial expansion overwhelmingly implied, hence essentially buttressing the heroic or benevolent myth of the expatriate Scot. Part of the introduction to the Canadian section of the MoS reads: 'Scots forged links with indigenous Canadians through trades and alliances. They also married into indigenous families and helped to establish a new nation – the *Métis*.' When enmity is invoked as in the depiction of the Red River settlers from Kildonan in 'Scots and the *métis*', it draws attention to the conflict of interest between the two main trading companies operating in the area – the Hudson's Bay and North West Companies – the Highlanders being 'caught up in the conflict'. At the MoI, inter-ethnic exchange and relations are also emblazoned in the information devoted to MacDonald descendants of the Chiefs of 'Glencoe and the Indians' – echoing Hunter's book title (1996). Yet it has been has shown that dispossessed Highlanders were no different from other European colonists and, in some cases, accompanied the expropriation of native inhabitants with ruthless persecution (Richards 1985: 269-71, Hunter 1994: 236-37, 1997).

Dovetailing the effect of the development of social history the 60s and 70s was a changing perception of museums, their social role and eventually their image. Originally they were regarded as institutions meant to collect, preserve and display collections; it was a view which underscored the primacy of 'objects' and the act of interpretation was little more than ancillary to the primary purpose of conservation. A concern for the public and its perception of museums critically altered the way museum practitioners approached their displays; the notions of communication, interaction and involvement became paramount and foregrounded the place of the 'subject' in museum work. If education had long been part of the agenda of museums (*Museums Journal* 1907: 43-44), communication and accessibility of knowledge or academic content were not necessarily explored or analysed in relation to museum collections and displays. This certainly changed and museums became seen as educators and communicators, as significant means of popularising culture and learning and of reaching out to ordinary people – an approach spearheaded by folk and open-air museums. Further, popularisation or democratisation of culture also encompassed the recreational dimension of museums, often overlooked in the past. Part of this change was to be found in wider social trends as people increasingly enjoyed more leisure time along with disposable income and educational patterns moved towards mass education. The reappraisal of the museum's societal function coupled with the need to attract a wider range of people to museum premises led the museum profession to much self-analysis and doubt. Frank Atkinson, Director of Beamish Open Air Museum, neatly encapsulated the dilemma that faced curators striving to shed the image of museums as fossilised institutions more akin to 'mausoleums':

> Do you wish to be serious institutions with little interest in the wider public and its needs – collecting and cataloguing and preserving your collections? Or do you wish to communicate with the untutored – but very experienced – public and interpret your collections for its greater enjoyment? I offer you the phrase enrichment through enjoyment (Atkinson 1975: 104).[6]

As museum professionals redefined their communication strategies and concentrated on the visitors' experience, the interpretation of the past moved away from artefacts to encompass different media or combination of media. In some venues objects became supplanted and exhibitions relied on audio-visual and graphic aids to provide the visitor with a multi-sensory experience. Attractive design and innovative displays became the trademark of a new breed of museums.

As already underlined few are the museums visited that support their presentation of the Clearances with artefacts; accounts are highly scripted, in some cases illustrated by paintings, sketches and maps or statistical information and, in three cases, by models reminiscent of Madame Tussaud's waxworks and exhibition methods. At Gunn HC, scenography, in the shape of the models of a '45 soldier, a shepherd and an emigrant woman, is unintrusive and is pared down to decorative rather than theatrical figures, acting as mere props (Figure 6.4).

By contrast, both *Aros* and Timespan used to base their presentations on contrived stage-craft. Timespan featured a scene re-enacting Patrick Sellar's brutal eviction of the family of William Chisholm which included an aged bed-ridden woman; in the sound-track, he could be heard barking over the crackling noise of fire the oft-quoted words: 'damn her, the old witch, she has lived long enough, let her burn', whilst *Aros* staged the intensity of exile with a departing ship and emigrants and numerous heart-rending accounts of contemporary eye-witnesses. It was history turned into spectacle performed through sets high on emotionalism verging on the dramatic and sensational for visitors inured to television images. This type of media-oriented displays which dramatise the past raised much concern among professionals who argued that when the presentation of history was hijacked by designer-led and consumer-centred displays targeting a media-demanding public, there was a danger of seeing the past turned into fantasy (Silverstone 1989: 138). Narratives of the past, by nature, often have a telescoping power on time, but when structured around climactic moments, they become a poor conduit for informed critical responses. It was an accusation often levelled

6 A decade earlier, Ian Finlay (1964), Director of the Royal Scottish Museum, had expressed a similar view.

against the proliferating heritage centres which drew on the public's interest in the past somewhat whetted by the production of TV historical dramas and documentaries. If, partly following new museological trends, heritage centres paid but scant attention to material culture, they were above all accused of trivialising and vulgarising the past through oversimplified storylines and dramatised replicas inducing a sense of vicarious experience (Davies 1989). The publicity brochures which used to advertise Timespan and *Aros* unambiguously foreshadowed the approach adopted. At Timespan, the visitor was invited to '*experience* our Highland Heritage ... Warriors and witches, maidens and murderers, prosperity and poverty, triumph and tragedy – the dramatic story of the Highlands' (my emphasis), while *Aros* was said to 'weave two historic threads together – the thread of life and the thread of hardship – in an exhibition that is dramatic, powerful and atmospheric' (Figures 6.5 and 6.6).

Figure 6.4 Clan Gunn Heritage Centre, Caithness
Source: Laurence Gouriévidis 2007.

Figure 6.5 Timespan's advertising brochure before its refurbishment, Helmsdale, Sutherland

Source: © Timespan.

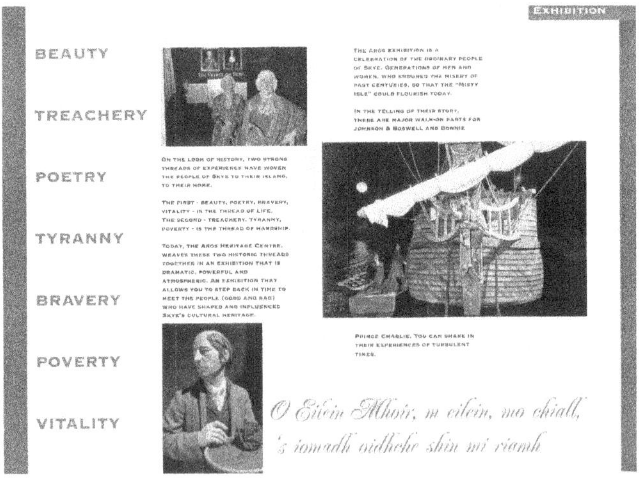

Figure 6.6 ***Aros*' advertising brochure before its refurbishment, Portree, Skye**

Source: © *Aros*.

Such an approach diluted not simply understanding of the complexity of the Clearances but also, to some degree, the educational function of such venues by privileging entertainment and affect over questioning and cognitive awareness; ultimately it smacked of condescension underestimating what visitors might be expecting. More importantly, it also tended to obfuscate an ideological message with important resonances in the present. As pointed at in previous chapters, Scottish heritage centres and museums, as opposed to what has often been reported about their counterparts south of the border, cannot be charged with presenting a sanitised version of the past, at least not with respect to the Clearances and the Land War; this, however, needs qualifying in view of the representation of wider conditions such as the concomitant process of colonisation, in which local emigrants were significant actors, and the pre-clearances era portrayed as tantamount to a lost golden age. Notwithstanding these reservations, representations of clearing policies are not devoid of pain and conflict – for a couple of venues social realism even meant positively wallowing in social distress and tensions – and echo the emblematic use of the Clearances in radical political discourse since the early twentieth century. The politicisation of the image of Highland wilderness/engineered emptiness may, however, well have been neutralised by dramaturgy in the museum context.

Towards a 'new museology'?

Irrespective of the misgivings expressed, the changes which affected museums and the way they 'package' their narrative underline their widening function within given communities, a function which is at the same time social, educational and also economic. The democratisation of museums embraced both public and producers, reception and production of interpretation. If widening the audience meant that museums dealt with issues and voices previously discounted, in terms of production, this meant involving local people in the elaboration of exhibitions. Many of the museums that emerged in the last quarter of the twentieth century were venues which engaged more actively with their surrounding community, for instance collaborating with schools and including outreach programmes (Walsh 1992: 161-62, Davis 1999: 52-62). Strathnaver museum is an early example of this as the Clearances exhibition was redesigned in the late eighties by local schoolchildren under the guidance of their teachers. The way Timespan has evolved since the late nineties is another significant illustration, incorporating in its premises an art gallery and, following the wishes of local people, a separate workshop. What the majority of the Highland and Island museums visited demonstrate is the significance of the involvement of civil society in cultural activities and more specifically here in the representation and telling of a past which focuses on the local and is consistent with the vision and aspirations of a particular community. More generally museums run by local trusts, be they historical societies or heritage groups testify to a movement which achieved greater scope in Britain roughly from the sixties and to museological developments which were witnessed worldwide. Hugues de Varine (2005a) who served ICOM as interim-Director and Director between 1964 and 1974 and now runs his own Communities and Development business in France, describes the change as a move 'from classical model building-collection-public (...) to another model of territory-heritage-population (community)' and summarises the wide-ranging causes of the movement:

> The worldwide earthquake of decolonisation and liberation wars, the international fora created by UNESCO and professional NGOs like ICOM, grassroots movements among ethnic minorities and local communities led, in less than one decade, to the emergence of a new generation of national and local cultural activists. They were interested in the preservation, use and transmission of their cultural and natural heritage, and were found in industrialised countries and in developing countries, in rural and urban areas. This had a strong impact on some young museum professionals, who were more interested in the ways that museums could be of service to society, the re-discovery of identity and independence rather that in "pure" scientific research. This resulted in the spontaneous movement which was later called "new museology", or even "ecomuseology".

Both the terms 'territory' contained in de Varine's alternative model or the label 'ecomuseology' also point towards an important societal development which specifically affected rural museums and the way they approached the representation of their locality and its surrounding: a greater public awareness of environmental concerns and the growth of a 'green' consciousness.

By the 1970s, one of the themes recurrently addressed by the Museums Association was the responsibility and role of museums in environmental conservation and education and the way environmental thinking was to be integrated in their field of activity (*MJ* 1972, Atkinson and Holton 1973). Open air, folk and rural museums were in the front line of such a programme and initiatives, linked as they were by the same thread: the study of man in his environment. The approach was seen as a means of breaking the mould of over-sentimental and nostalgic rusticity so often typifying their exhibitions. As Fenton (1973b) observed:

> We are accustomed to see aspects of man's cultural and natural environment in museums, but for the most part what is presented is a frozen point in time. Rarely is an attempt made to show processes. Changes is part of both human and natural history, the interaction of one on the other conditions environment to a quite considerable extent ...[7]

Ingliston's Scottish Agricultural Museum, situated on the outskirts of Edinburgh, was established with this vision in mind, enabling an increasingly urbanised population to understand its rural environment (Fenton 1978: 168). Once moved to

7 This article is reprinted in Fenton (1985: 34-42).

Kittochside and complemented with a working farm, the museum retained the same ethos adding to its exhibition the social history section in which the Clearances are now found. More generally, the 'greening of museums'[8] was advocated and tapped into from a variety of angles, reflected in the way museums were to incorporate the issue in their displays – sometimes simply paying lip service to the theme, particularly when doing little to connect past actions and ways of using the land to present features in the landscape. The social potential of such an approach was advanced by some who saw it as widening the function of the museum and therefore attracting community interest and participation (Cubbon 1972: 96, Atkinson and Holton 1973: 141). This was also combined with an economic argument given the fashionable appeal of environmental conservation in the public – a means of popularising the museum. The economic argument certainly had its attraction at a time of increased economic stringency. The so-called wild landscape of the Highlands and Islands, bearing the archaeological and ecological footprint of past human activities was ideally placed to respond to the development of such an approach, and the traces of past human habitations, field systems and settlements characteristic of the pre-clearance era, offered enormous possibilities, particularly for visiting tourists looking to blend appreciation of the scenery and cultural understanding with mild exercise. As one cultural actor commented:

> A museum in a rural area might be in a good position to provide better interpretation of the environment for visiting tourists. Many rural museums already have displays about the local landscape. Weather permitting, tourists are often looking for a short half-day trip that enables them to visit important local sites for themselves (Hillier 1994: 16).

Clearances trails in particular, in no small measure, also fulfilled an important economic role, providing tourist with an 'outdoor activity'.

The social role of museums is, in many cases, tightly knit with their economic dimension, yet, at times, it seems that museums are not so much emblematic of the increasing importance of civil society but of the commercial sphere, somewhat sacrificing their pedagogical and social soul on the altar of commercialism and populism.

8 The expression is borrowed from the title of a conference organized by the SMC in November 1992, see Ambrose (1992b).

Chapter 7
The Economics of Clearance Heritage

Our past has never been so popular. Nor has it to date played so important a role in the leisure and tourism economy (Ambrose 1989a: Preface).[1]

By the 1980s the rapid growth in the number of museums and their changing character raised concerns among professionals not simply with respect to the resulting duplication and competition for cash and visitors, but above all as regards the very image and ethics of museums; such doubts were testimony to the profession's crisis of identity and self-definition:

> That certainty about the museum's separation from the market place, and of its educational as opposed to entertainment role is in crisis. Not only has the museum shop expanded but all areas of museums show a growing influence of commercial models, while museums themselves are "sold" to the public as never before (Lumley 1987: 82).

Although many venues in Scotland as elsewhere were reluctant to embrace wholeheartedly the commercial sphere with its emphasis on customer care, strategic planning, market research and product analysis, others like Timespan and later on *Aros*, applied new concepts and sought to tailor their 'product' to a design-aware public looking for entertainment along with information, providing their visitors, through multi-media technology, with a sense of the past, place and community they were discovering. When, in 1990, Timespan received the newly created audio-visual prize, part of the Scottish Museum of the Year Award, the announcement reflected the sense of flux which characterised the profession, insisting as it did on the diversity of the winners, 'typical of the range of Scotland's museums, from noisy working machinery at Coatbridge, to quiet scholarly research in Wester Ross, with a lot of fun in between at Helmsdale …'. Yet, Timespan in spite of its award was also described as 'perhaps more a heritage centre than a museum, evok[ing] some of the more colourful events in the Highlands vividly in sounds and images' (*SMN* 1990: 11). A history teacher, visiting Timespan from Paisley during its first season, enthused in the pages of the local paper commenting that he was most impressed, the centre being 'very much a "People's Museum" in the modern style' (*NT* 1987c). Timespan and *Aros* are the most powerful illustrations of the influence of the market place and populism on the representation of the past

---

1 Timothy Ambrose was Director of the Scottish Museums Council from 1986 to 1994. The papers in this collection were given during a conference organised by the Scottish Museums Council in November 1988.

and place to visitors. They both tapped into the public's appetite for heritage and seem to be fitting examples of the 'nostalgia business' showing that exhibiting the 'drama' of history pays dividends by touching the quick of audiences, be they related or not to the region. However, the way they later evolved is also indicative of their differing emphases and priorities – their core values.

The purpose of this chapter is not simply to consider whether and in what way appealing to specific 'market segments' or heritage consumers affected Clearance narratives, but also to examine the development of clearance-related heritage in the context of strategies of economic development in the highland and island region, paying particular attention to the place of museum spaces in the leisure and tourism industry. Such strategies have increasingly been associated with the notions of social empowerment and fostering of community belonging and participation, demonstrating that the economics of heritage are not easily distinguished from its social potential in the eyes of many funding agencies and local actors.

Museums and economics

By the time Timespan became an award-winning museum, the economic benefits of museums at a local, regional and national level had been well rehearsed by the SMC and many other agencies, whilst museums themselves increasingly had to justify their existence in terms of customer appeal and value for money, as their paymasters, such as local authorities, became under pressure to contain expenditure and increase income. The language of straight commercialism pervaded the pages of the *Museums Journal* and *Scottish Museums News* with courses and seminars on marketing regularly publicised; also emblematic of the era were the central themes of the annual conferences of the Museums Association: 'financing museums' in 1985 and 'marketing' in 1988. Museums were widely recognised as pivotal elements of one of the major props of the Scottish, let alone British economy, the leisure and tourist industry. They were encouraged to attract funding from a variety of agencies and form partnerships – as seen earlier in Chapter 3 – but the fear was that this could lead to an anarchic situation as project developers took advantage of the heritage craze and jostled to climb on its bandwagon. Partly aimed at promoting greater co-operation between public bodies and museums, the Association for Scottish Visitor Attractions (ASVA) was established in May 1988. The definition it provided for 'visitor attractions' is worth quoting:

> An established excursion destination, a primary purpose of which is to allow public access for entertainment, interest or education; rather than being solely a retail outlet or venue for sporting, or film performances. It must be open to the public, for published periods each year, and should be capable of attracting tourists or day visitors, as well as local residents (Hayes 1991: 12).

It was a definition broad enough to accommodate establishments with diverging priorities – from those targeting tourism business to those collecting and preserving collections; interestingly, topping the list of attractions visited in 1989 were museums and galleries, before historic buildings and ancient monuments and, in third position, heritage and visitor centres (ibid.). The development of museums – seen as enhancing the visitor's experience – was recurrently analysed in tourism research and one of the benchmarks used to evaluate the performance of a tourist season was the number of visitors selected attractions received. Focusing on the nature of the appeal, history and countryside were presented as the 'principle motivators of international tourism', more so in Scotland than anywhere in the UK (Borley 1988: 16, Scottish Tourist Board 1998); these factors were not taken lightly at a time when it was also predicted that, by the end of the millennium, as many as one billion people could be on the move internationally every year (Ambrose 1990b). In these two areas, history and countryside, the Highlands and Islands had much to offer and the interpretation of the Clearances snugly fitted the bill, blending human and environmental concerns.

Changes in the museum world were signs of wider social and economic trends in Scotland and the UK at large. The growth of leisure and tourism had long been forecast; by the sixties, it was expected that people would take more than one holiday a year and that car ownership would extend to lower socio-economic groups as people enjoyed more disposable income. But the importance of the tourist industry in the nation's economic base grew as successive UK governments embarked on de-industrialisation and favoured the growth of services in the eighties and nineties. Not only could museums generate jobs in the context of soaring unemployment figures and economic re-structuring, but they could also, with the provision of improved cultural infrastructures, contribute to the regeneration of an area by drawing inward investment and encouraging firms to relocate. With the new political orthodoxy of anti-corporatism and reduced state subsidies giving pride of place to free market and competition, museums had to diversify their sources of funding and think of visitors as 'customers' or 'consumers', undertaking surveys and research to assess their responses and anticipate their wishes. More recently with the advent of IT and web curation, museums have been able to advertise their collections as never before and export heritage world-wide, a valuable asset at a time of fast-increasing intercontinental travel.

Local museums and tourism

For the Highlands and Islands, more perhaps than elsewhere in Scotland, the value of tourism to the regional economy had long been recognised. When the Highlands and Islands Development Board was established in 1965, the region was branded a depressed or problem zone of high unemployment, population loss and economic fragility. Tourism, because of its ability to generate income and employment even in the remotest areas such as the Western and Northern Isles, and in accordance with the guidelines contained in the White Paper on the Scottish Economy, was classed as a potential growth point, 'the second great prop available to us … capable of making a rapid impact on the [local] economy' (HIDB *Annual Report 1966*: 3, 23-27) – in spite of its seasonality and vulnerability to external factors. In 1975, income derived from tourism was already estimated between £100-£130m or roughly 25 per cent of the 'Gross Highland product', a significant proportion of which was expended by overseas visitors making them a valuable target. As regards employment, tourism provided jobs for over 11,000 people (HIDB 1979: 7.5 and 7.6). Over 15 years later, in the early 1990s, some 20,000 people, it was reckoned, found full or part-time employment in tourism or 20 per cent of the local workforce. Tourism was said to generate over £500m annually, amounting to 20 per cent of the area's gross domestic product (HIE *Annual Report 1991-92*: 36, *1992-93*: 38) and to be 'worth more in financial terms than the area's fishing, fish farming and livestock industries combined' (ibid. *1991-92*: 36). In the Western Isles, 'the most fragile economy' in the region in the late eighties, to combat unemployment and diversify the economy, it was again tourism that was earmarked with a special local tourism development programme – marketed as Crofting Life Holidays (HIDB *Annual Report 1989*: iii, 26). As traditional industries declined and the hopes raised in the early seventies by the oil industry failed to materialise evenly, tourism seemed to offer more consistent prospects of growth and more geographically widespread benefits. However, it was and still is also prone to fluctuations and extremely susceptible to external influences ranging from poor weather to a variety of factors – national and international – such as unfavourable exchange rates for overseas visitors, oil recession, fear of terrorism, and even ecological and environmental disasters as with Chernobyl or foot and mouth disease. Reaching a peak in 1973, tourism in the Highlands and Islands then stagnated, to finally spiral downwards, and the 1982 report of the HIDB makes gloomy reading outlining the declining trend of the previous years. Thereafter it gradually recovered but it was a while before it was to match its mid-seventies performance, whilst still suffering periodic glitches. By the turn of the century, it was one of the area's recognised assets in an economy that tended towards increased diversification; this was a time when sombre or qualified appraisals were discarded in favour of optimistic grammars emphasising confidence, enterprise and achievements; the label 'highland problem', it was claimed, could finally be consigned to the past (HIE *Annual Report 1996-97*: iii): 'the area's fortunes [had] turned around remarkably' (HIE *Annual Report 2000-01*: 8).

It therefore comes as little surprise that, as part of the Board's early strategy to develop tourism, one particular focus was to increase facilities – particularly 'wet weather facilities', also favoured by the STB – in order to provide visitors with activities retaining them in the region and dispersing them more widely across the area (HIDB 1981, Museums and Gallery Commission 1986: 110). In the early seventies, the Strath of Kildonan was the object of a specific report which, as a means of boosting tourism in the locality, advocated the establishment of a small museum 'where visitors could learn of the Clearances and the gold rush of the last century ... [It] could be combined with a craft shop and a satellite information centre to make a viable economic unit' (HIDB 1970: 51). Incidentally both HIE and STB emphasise the need for a gift shop to 'enhance the visitors' experience' to museums (survey answer: Mallaig HC) and top up income. The report seems to have had few tangible effects and it took some time for the Kildonan project to take shape, a project which relied on the sustained efforts of local enthusiasts; in July 1987, Robert Cowan, the Board's Chairman, officially opened Timespan whose funding came from a combination of sources, public and private, including the Highland Regional Council (£62,000) and the HIDB (£44,000) (HIDB *Annual Report 1987*: 20, *NT* 1987a, 1987b). Slightly later, a similar funding pattern applied to *Aros* for which the Local Enterprise Company – after the re-structuring of the HIBD into HIE – contributed some £375,000 (HIE *Annual Report 1992-93*: 26). *Aros*' high-tech exhibition was a costly venture and its funding structure, using a high proportion of public money, attracted criticism on the grounds that it was likely to meet the fate of many similar heritage attractions elsewhere in the UK, which were starting to close down at a fast rate because of competition and lack of customers (*WHFP* 1993b).

By contrast, with low maintenance costs but fulfilling the objective of retaining and even piloting tourists within an area, the creation of trails and installation of interpretive panels also became favoured by various agencies. It was another means of enhancing the visitors' experience by drawing their attention to the region's cultural – and environmental – wealth as well as inviting further exploration. In its 1990 annual report, HIDB announced that its 'aim [was] to create a series of themed trails which will increase interest and encourage tourist to travel and stay in some of our more remote places' (1990: 30). It was a programme strongly supported by ASVA which reckoned that:

> In some places a striking feature is crying out for interpretation: a glacial terrace, a hill fort or a ruined castle. Less obvious features, but of equal potential interest, might be traces of shielings, the track of an old drove road or the outline of a long deserted township (Association for Scottish Visitor Attractions 1991: 1).

Given the fact that the profile of visitors to the Highlands and Islands was deemed 'relatively well educated and informed' (Westbrook et al. 1999: 15), both museums and heritage trails seemed to provide valuable opportunities, the former offering a wet weather option, the latter a complementary development, increasingly promoted and advocated in a heritage market that threatened to become museum-saturated (ibid. 45).

Once up and running, the vast majority of venues surveyed rely heavily on admission fees and retail to survive and, in some instances, it makes little doubt that narrative and museological choices were strongly directed by the need to appeal to visitors/customers. 'The Cartland saga', as the *Northern Times* branded the debate which erupted as a result of the addition of a Barbara Cartland room, part of the second phase of Timespan's initial exhibition, provides anecdotal evidence to the tensions faced by heritage practitioners (*NT* 1988c). Barbara Cartland married into the McCorquodale printing family who have owned an estate in the Strath of Kildonan since the Twenties; she holidayed in the Strath regularly in the summer and, on her own admission, used the region as a backdrop to a number of her books (*NT* 1988b). One local councillor regarded the 'pink' room as 'a damn disgrace', an incongruous appendage in an exhibition devoted to the life and history of the locality, whilst Robert Cowan, again invited to perform the second opening, was dismissive of such censure, declaring:

> Timespan is a very good example of what can be done with history – you have succeeded here with making history fun. There seems to have been a little bit of a "stooshie" on the subject of Barbara Cartland; I don't think we should object too seriously to a little bit of modern history … and I am sure that Barbara Cartland will help bring a bit of publicity and numbers to the centre (*NT* 1988a).

Unfortunately, the pink room somewhat took the shine of the newly installed Clearances display – also introduced during this second phase – which did not even find a mention in the local press, bearing out the point made by disgruntled locals.

One of the areas in which these venues can easily be differentiated is their dedication to the sphere of cultural heritage over their lifespan. Using marketing-speak, all are to some extent sensitive to their 'product life cycle', yet the strategy they follow to adapt to evolving conditions might be more driven by the market than socio-cultural aspirations and the promotion of local history, bringing out the core values to which they adhere. For the majority, 'developing or consolidating the product' meant improving form or layout such as including an audio-visual presentation (Mallaig HC), a Clearances section (Arran HM) or refurbishing an exhibition with more design-conscious displays (MoI and Ullapool M&VC) or, even more tellingly as regards museological practice, safer and better controlled equipment to accommodate loans of artefacts from the national museums (Timespan); for the purpose of research, facilities could be developed or a library or archives could be added to the premises (MoI, Strathnaver Museum, Arran HM, Glengarry HC, Timespan) – such choices confirming their cultural orientation. For one, *Aros*, it meant changing emphasis and jettisoning local history in favour of the more fashionable theme of environment, the argument being that the initial exhibition, after a few years, did not succeed in attracting repeat visits.[2] Although *Aros*' web page continued to advertise a historical dimension – 'a small exhibition dealing with the Clearances and the period of the Jacobite revolution',[3] a visitor in search of historical heritage can only be sorely disappointed as the display is a mere shadow of its former self and provides but the sketchiest of historical backgrounds, even more starkly articulated around the most 'dramatic and powerful' elements of

2 Personal communication, June 2007.

3 http://www.Aros.co.uk/skye-the-island.html [accessed 18 June 2007 and 10 June 2008].

the area's history – to quote *Aros*' initial promotional leaflet. *Aros*' subtitle was also amended and the centre no longer stands as 'Dualchas an Eilean' which suggested a specific vision of Skye's heritage and culture.

Figure 7.1 ***Aros*' advertising brochures before and after refurbishment in 2002**
Source: © *Aros*.

Whilst it could appear to be a glaring instance of the heritage business gone badly astray, sacrificing historical integrity to financial motives, in commercial terms, *Aros*' re-orientation also signifies the entrepreneurial vitality and 'flair' of a local business, at best adapting to societal shifts, at worst cashing in on the latest customers' fad.

Cultural and ancestral tourism

Aros' re-invention and the environmental component it already featured when first opened, are also evidence of another significant resource that agencies concerned with tourism development have been keen to exploit – the region's natural environment. The landscape of the Highlands and Islands has long held a powerful fascination for travellers (Smout 1983) but, with the growth of ecological consciousness, the region was deemed 'well-placed to capitalise' on the public's environmental interest, bracketed 'green tourism' (HIDB *Annual Report 1990*: 32). From this perspective, museums presenting the history of the region have much to contribute and narratives of the Clearances provide a meaningful historical context in venues attempting to interpret the making of the region's landscape and its apparent wilderness – the interaction between man and the environment over time. Such is the case of the displays at Strathnaver Museum and Arran HM, which address population distribution and decline alongside new forms of land use. In the specially-commissioned videos and slide shows at Timespan and Ullapool M&VC, a similar perspective is adopted. Whilst, at the latter, *The People of the Loch: The Lochbroom Story* fuses natural and human history, at Timespan, attention was called to the ecological impact of sheepgrazing.

Growing equally critical, in marketing terms, to the region's appeal has been its Gaelic heritage. In the late seventies, however, this was not altogether a matter for overall consensus and Gaelic culture and tourism were seen as uneasy bedfellows by some who were concerned that the tourist industry, in spite of its benefits to the local economy, could also, owing to its seasonality, inhibit long-term economic and cultural vigour, in particular in the west and the islands.[4] Incidentally, tourism read as detrimental to the area's cultural heritage was an argument still used by *Aros*' detractors in the early 1990s (*WHFP* 1993b). Nonetheless, by the mid-eighties, Gaelic heritage was envisaged as a critical tourism asset; it was discussed, for instance, by the Highland and Island Tourism Council during one of its meetings in 1986 and thereafter became a prominent feature of HIDB's and HIE's reports. Fostered by the cultural renaissance, the growing strength of Gaelic was interpreted as an eloquent signal of community's self-confidence and used as a motor for socio-economic regeneration (HIDB *Annual Report 1986*: 17, *1987*: 31, Pedersen 1995). It was also fuelled by a supranational vision emerging from Europe, which sought to give greater recognition to the distinctive regions and minority cultures within the Union. This followed, in particular, the adoption of the new Treaty on European Union (Maastricht Treaty) in February 1992 and the establishment of the Committee of Regions, which gave pride of place to the principle of subsidiarity. All in all, the 'Gaelic tourism product' could profit from Gaelic interests at home and overseas, comprising, for instance, tourists from the Celtic areas of Europe, from other bilingual or multi-lingual areas in the world such as Canada and, of course, those tracing their ancestry to Gaelic-speaking Scotland (Pedersen 1995).[5] All Skye museums, although very feebly in the Museum of the Isles, raise the question of cultural and linguistic distinctiveness emphasising the eventual resilience and richness of the island's heritage in narratives depicting the Clearances not simply in view of their resulting decline in the local Gaelic-speaking population, but also for the long-lasting scar they left in the Gaelic-speaking community's psyche and

4 The position and its effect on the HIDB is described in Grassie (1983: 37-8).

5 When this paper was published, Roy Pedersen was Development Director with *Commun na Gàidhlig*, formerly Head of Social Development with HIE.

self-worth. When first opened, *Aros*, was singled out by HIE – relayed in the press – as a successful centre interpreting the Gaelic and cultural dimension of Skye to visitors (HIE *Annual Report 1992-93*: 39, Calder 1993); it certainly succeeded in marketing itself as such over time, its Director declaring in 1997: '*Aros* is much more than just a heritage visitor centre and has played a real role, since it was established five years ago, in the community and the development of the Gaelic language and associated culture' (*Scottish Daily Record* 1997). It was also given prominence in museological studies highlighting the active involvement of local people reclaiming their past and identity through heritage (S. Macdonald 1997) – a feature it shared with other venues in the Highlands and Islands and which was no novelty in the museum representations of the region's heritage. On Harris, at *Seallam!* VC, where the visitor is also introduced to Gaelic's rich oral tradition, its 'story' is not so much inserted in a narrative of struggle against suppression, but liveliness in the outer isles and legacy across the Atlantic. Similarly, the HFM under the auspices of the Highland Council, was keen to incorporate Gaelic in its interpretive 'living history' approach and called on the expertise of venues with similar programmes in the pages of *Museum International* (Menabney 2001). Perhaps more symbolically Gaelic also appears on the panels of a few venues – and trails – often being restricted to titles, such as MoI and Mallaig HC – for the cynics, a token effort at inclusiveness, adding a sheen of local colour, or a means of ticking another box in the check list of potential funders.

Within cultural tourism, the increasing emphasis laid on emigrants and their destinations in Clearance narratives also bespeaks another feature of the region's heritage harnessed for a very distinct – and lucrative – niche; it is a 'tourism product' which responds to the anticipated needs of a specific market segment: the great reservoir of genealogical visitors in search of their roots, that descendants of clearances – and other Scottish – emigrants in the United States, Canada, Australia and New Zealand constitute. The words of David Lowenthal effectively encapsulates the diasporic experience of return tourism: 'Massive migration sharpens nostalgia ... Diaspora are notably heritage hungry' (Lowenthal 1996: 9). Satisfying this appetite was one of the objectives stated in the 1981 report, prepared for the HIDB, which defined the type of tourist facilities required in the Highlands and Islands:

> One of ideas that came to the fore during the "brainstorming session" was that of capitalising on the growing interest of genealogy ... Whether this qualifies as a wet weather facility is perhaps open to doubt, but the concept was too attractive to ignore (HIDB 1981: 18).[6]

The same idea underpinned the creation of *Co Leis Thu?* which has offered this type of resource in the outer Hebrides since it opened in the early 1990s. It is now

6 Genealogical tourism is also an opportunity mentioned in the SWOT analysis of the independent museums sector in Westbrook et al. (1999: 26, 43).

the genealogical arm of the complex clustering around *Seallam!* VC, and includes a visitor and art exhibition centre. The genealogical service is run along commercial lines with some information available on the internet. Among the other venues surveyed, some, such as Glengarry, attribute the initial motivation behind their creation to demand from overseas descendants of emigrants in Glengarry County, Ontario and in the antipodes, and, unsurprisingly, the discovery of the cleared village of Daingean was advertised in the 'diasporic' Scottish press (*Scottish Banner* 1999).[7] Understandably, root tourism is the cornerstone of Clan Museums – including Strathnaver Museum serving Clan Mackay – which, not only seek to provide clan members with genealogical resources, but also with a focal point for clan history and activities, in particular gathering. Overall, in Clearance narratives, the more extensive treatment of the nexus eviction and emigration might partly be explained by the need to respond to a quest for origins and expatriate identity on the part of descendants of cleared Highlanders and Islanders anxious to revisit their ancestral roots and a shared memory of exile. Their desire might be exacerbated by the nature of the societies they live in – multi-cultural and multi-ethnic – striving to acknowledge the distinct experiences of the diverse communities composing them. Moreover, the way Clan museums weave the notion of victimhood into their narratives of the Clearances (see Chapter 5) owes much to a need to buttress unity and a global sense of kinship and goes a long way towards accounting for the silences and narratorial shifts found at the MoI, where patronymics associated with the clan do not explicitly feature alongside perpetrators of evictions, but alongside evictees. The creation of the museum was an integral part of the development plans of the Clan Donald Lands Trust, when it was established in 1971, and clan members worldwide have been instrumental in the founding of the Trust and its operations – for instance, appeals were made to members for the initial purchase of the land. Additionally, as part of its funding, the museum also receives a grant from a US Trust, the Glencoe Foundation. Its exhibition was set up by the archivist with approval and input of the Director of Trustees and in texture the narrative may be said to reflect what a recent Hebridean emigrant to Canada underlined: 'As an expatriate Gael, I carry the additional baggage of the narcissism that urges one to transform one's identity into a self-pitying sense of victimisation' (J.A. MacPherson 1995: 279). The way in which the MoS has constructed its Clearance narrative seems to take stock of this attitude. Heritage tourism can have a variety of effects on the narratives produced and their modulation, the memorialisation of the past becoming prey to the collective imaginaries held – or thought to be held – by the group targeted. The example of the commemoration of the Second World War on the Solomon Islands has been used by Ashplant et al. (2004b: 84, note 255) to illustrate the point that 'the imperative towards tourism-development may undercut, rather than further, the process of questioning memories constructed from above'. In the case of Clearance narratives, heritage tends to further the

7 The paper describes itself as 'the largest Scottish newspaper in the world outside Scotland'.

people's memories of the process rather than elite memories; furthermore, in a large measure, it is the memory of a victimised group.

As the MoI exemplifies, another aspect also stressed in relation to descendants of emigrants is the fund-raising opportunities that the 'wealthy and influential overseas diaspora' offers, notably in relation to the development of cultural tourism projects similar to Clan Donald Centre (Pedersen 1995: 295-96). Likewise, the team planning for the MoS tried to capitalise on this market in its fund-raising activities, which extended beyond Scottish shores to embrace places where Scottish emigrants had settled. It homed in on clan and St Andrew's societies and was eventually successful in matching the funds awarded by the Heritage Lottery especially through its American appeal; the American – as opposed to the Australian – appeal was particularly rewarding, not simply because Scottish descendants had been enjoying a renaissance and were about to be given a higher profile through the official designation of Tartan Day, but because fundraising was undertaken through effective local presence (McKean 2000: 91). Just as clan museums act as points of convergence for their scattered members, so does the MoS for the Scottish diaspora, most obviously through the section honouring Scots abroad, 'Scotland and the world', whose video starts with the arresting statistics: 'there are 25 million people round the world who claim Scottish descent. Just five million live in Scotland, how did this come to be?'

So far the economics of heritage has been envisaged in terms of tourism or visitors targeted: another angle from which it has to be appraised, concerns its source or anchoring within a given population and environment in the Highland and Island region and its impact within that community. The economic spin-off generated by tourism is often closely entwined with local development and social dynamism.

Heritage and local development

The socio-economic impact of museum is largely assessed in terms of their function as tourism attractions, however their value to local communities' sense of self and place in contemporary society is another aspect that funders came to appraise. But what of museum actors themselves, how do they assess their role? Answers to the question related to the museums' main objectives, in the survey addressed to independent museums, provide valuable insight into their perceived function within specific communities (see Appendix, section 7). The breakdown of answers received for that question is recorded in Table 7.1. Of the ten respondents, all listed 'tourist attraction' and 'teaching or informing visitors about Highland and Island history'; more precisely, among the six who did rank their answers, 'tourist attraction' was never a primary objective with one respondent giving it second place and the other five rating it third or more, whereas 'teaching … history' came first with four of them, the other two placing it in third and fourth position. Some of these exhibitions are not solely concerned with history and their broader multidisciplinary approaches is reflected in the number of positive answers concerning the 'dissemination of Highland and Island culture'. Importantly, nine of the respondents placed 'resource centre' as a significant function, five of them ranking it in second and third position, pointing to the fact that in many remote areas such venues double up, for example, as archives or genealogical centres (see Table P.1). Only one mentioned profit and half 'contribution to the local economy', partly because the question narrowed the understanding to employment, discounting other upshots – such as the impact of tourism – and, as has already been seen, the majority of independent museums are staffed by volunteers. Expanding on this aspect, a handful provided further information specifying the number of staff they employed, casting further light on the seasonality and short-term nature of the employment on offer. For instance Gunn HC offers one summer job; other venues, part of larger complexes, included jobs made available overall – Clan Donald Centre as a whole offering 15 year-round jobs and many seasonal. Similarly, at *Aros*, the staffing situation in 1996 was 23 full and part-time staff, twelve of which were on a year-round basis (*WHFP* 1996c). However, since survey answers reflect a current situation, they do not provide any inkling about the past. Little is revealed about previous conditions which did enable these small museums to offer other employment opportunities, in particular through the various training programmes which the Manpower Service Commission (MSC) ran, until it was disbanded in 1988. Jobs ranged from restoration and construction work to documentation, and were of a short-term nature. The Arran HM, in 1978, produced a job creation scheme for the renovation of the stable block which was to enlarge the exhibition (Small 2002) and, in 1987, the MSC provided funding so that the museum could employ three people for 52 weeks to create a new reception area, pay-point and shop in the roadside cottage previously used as a display area. In 1976, Farr Church which was to house Strathnaver Museum, was restored through a grant given to Caithness District Council by MSC to employ four workers – a decision reported in the press as 'history creates jobs for future' (D.MacLeod 1976); in the 1970s and 1980s, Pat Rudie, the then museum's curator provided trainees with museum skills such as recording and cataloguing. For Timespan, the involvement of the MSC was mainly restricted to landscaping the entrance area.

Table 7.1 Breakdown of survey answers to question: 'What are the main objectives of the museum/centre?'

	SMoIL	MoI	Arran HM	Strathnaver Museum	Gunn HC	Timespan	*Seallam!* VC	Mallaig HC	Ullapool M&VC	Glengarry HC	Total number of answers
Contribution to the dissemination of highland and island culture	✓	✓		1	4	✓	2		4	✓	**8**
Teaching or informing visitors about highland and island history	✓	✓	1	3	1	✓	4	1	1	✓	**10**
Make a profit					5						**1**
Contribution to the local economy		✓			6	✓	3		5		**5**
Resource centre		✓	3	2	2	✓	1	2	2	✓	**9**
Tourist attraction	✓	✓	2	4	3	✓	✓	3	3	✓	**10**

Table 7.2 Breakdown of survey answers to question: 'Can the museum/centre be described as a museum/centre of the local community?'

	SMoIL	MoI	Arran HM	Strathnaver Museum	Gunn HC	Timespan	*Seallam!* VC	Mallaig HC	Ullapool M& VC	Glengarry HC
Involvement with schools		Envisaged	✓	✓		✓	✓	✓	✓	Sought
Temporary community exhibitions				✓		✓	✓	✓	✓	✓
Involvement with local historical society							✓			✓

As already stressed in this chapter, economic outcomes spill over the perimeter of the museum, a point underscored by Timespan's respondent who added that the centre was 'conceived at a time of depopulation of Helmsdale with a view to providing employment within it and spin-off for hotels and B&B establishments and shops etc.' On its opening, the centre was commended for reviving not simply an old derelict building, but the whole village with the many benefits yet to come (*NT* 1987a). The article is in stark contrast with the way 'untidy' Helmsdale was catching the headlines ten years earlier, with plans to 'give it priority facelift' reported (*NT* 1976a). Timespan, like many of the venues created in the late 1980s and 1990s, point to the connection between museum and development related to economics and apprehended in terms of job creation, training and acquisition of skills, as well as value added to a place. As heritage critics highlighted, the heritage movement in the region, as elsewhere, was invigorated by the existence of a variety of funding sources, and venues, cashing in on this golden opportunity were, more often than not, spawned in areas of economic decline and seen as potential panacea for local malaise, often losing sight of their *raison d'être* after a few years. *Aros*, as seen previously, is the most conspicuous example. Three years after its inauguration, it was said to have been 'developed on the basis of telling the story of Skye and its people. Now, it washes its face by having a large restaurant, café, and shop' (I. Sutherland 1996).[8] Ironically, when at the same period, its owners were applying for further funding to the HLF to build an auditorium, its plans were initially opposed by Portree Community Centre Association, fearing that they might jeopardise the chances of further local funding bids. The arguments used hinged on *Aros* as a privately-run tourist amenity with retailing and catering space and its heritage dimension did not even receive a mention (*WHFP* 1995, 1996a, 1996b).

Yet, the contribution of heritage venues to local development also extends beyond economic assessments to encompass the social and cultural spheres and this became part of the orthodoxy which governed much the funding for the establishment of new ventures from the late eighties, as museums grew to be associated with the enhancement of a living environment and regarded as foci for the provision of a diversity of cultural activities within a community. They were also envisaged as dynamic forces fostering initiative, participation and ultimately injecting self-confidence within the communities they served. The potential of culture in social and economic regeneration, in revitalising depressed and under-performing communities became part of political discourse and stimulated policies and action. HIDB reports are good indicators of this changing cultural focus; 'language revival' was one instance, as it came to be regarded 'as a powerful motor for both social enrichment and economic regeneration' (HIDB *Annual Report 1987*: 31) and became therefore worthy of support. More widely, people's and communities' pride in their own cultural identity was thought to be the key to

8 The journalist was quoting Bill Taylor, then Project Officer for the Highland Interpretive Strategy Project.

greater confidence, 'generating activity and winning economic prosperity' (HIE *Annual Report 1995-96*: 39). Originally framed as 'social development', in the late eighties the promotion of culture was styled 'community action' and from the mid-nineties 'strengthening communities' with their emphasis on grass-root initiatives.[9] The labelling shift, which duplicated Euro speak, also signalled the deeper ideological shift that the Thatcher era heralded, placing a high premium on proactive groups and self-help projects for which government agencies and quangos would simply act as 'enabling' instruments. In this sphere Highland and Island museums were ideally placed as they were the fruits of the persistent and determined endeavour of local individuals and groups; the tradition of volunteering, independent effort and resourcefulness was to serve the local museum sector well as it gained momentum in the 1980s. Additionally, within the province of culture, museums were ideal loci where cultural awareness, pride in one's place and past could be fostered and the *Museums Journal* recurrently featured articles in which museums were shown as fronting the campaign against urban disintegration and social exclusion in town and country, through the encouragement of community participation, meeting and projects.[10] In this area again, many of the early independent Highland and Island museums surveyed seemed, by their very nature, to be precursors, enlisting the support of local resident volunteers for their development and survival; in 1978 the restoration of the cottage at Arran HM was performed by volunteers. Later venues, such as Timespan also relied heavily on the knowledge and skills of locals for the construction of exhibits such as its croft house or village shop. Cooperation with educational institutions is another key area of social development which museums were encouraged to develop. From the inception of Strathnaver Museum, the collaboration with local schools was, and still remains, the trademark of much of its exhibition and the panels, paintings and many of the models supporting the Clearance narrative were produced by local secondary schoolchildren under the supervision of their art and history teachers. For later museums, opened in the 1990s, working with schools and holding temporary community exhibitions was to become standard practice conditioning grant-aid; 'multi-functionality' was seen as an asset turning museums into cultural focal points in small dispersed communities (Broad Oak Consultancy 1993: xiii). Table 7.2 summarises the answers provided to section 9 of the survey related to social impact. These answers, combined with those concerning other services provided, explain why many respondents insisted on the multi-facetted role of their venues. For *Seallam!* VC, which also acts as a learning centre for the UHI, it was specified that the aim was to give visitors and locals as wide a range of facilities as possible in a remote area. Significantly, the history exhibition at Glengarry HC was initially the core of a small visitor centre, housed in a 'Canadian-style' log cabin, before it

9 See the analysis of the strategic aims of such funders as HIE/LEC and the Highland Council in Westbrook et al. (1999: 27)

10 See for instance: Lorente (1995), Carrington (1999), Sandell (1999), Newman (2001), Heywood (2002).

was moved to a multi-purpose building including research and library resources; it was described in survey answers as 'an integral part of a community centre offering facilities for concerts, conferences and seminars', which, it was hoped, might soon be complemented by an internet café. Timespan is another example of a venture initially conceived and managed as a visitor attraction, but which, over time, developed into a cultural centre and community base with space gradually made available for an art gallery, archives and for the use of local people; oral history events, craft or photo nights are recurrently held in a workshop which locals had requested, when plans for the renovation of the centre were being discussed.[11] In many of the venues – particularly those run by trusts – exhibitions have become one facet of a larger kaleidoscope, in which the 'tourist' is no longer the only focus nor beneficiary.

Hugues de Varine, in a recent essay, based on his international experience and considering the influence of heritage on local development, states that such development cannot occur without the genuine, active and conscious participation of the community which is the guardian of its heritage (Varine 2005b: 15). Certainly, in the Highlands and Islands, a large number of the heritage venues considered can be regarded as marks of the initiative, creation, inventiveness, not to mention entrepreneurial drive of members of their surrounding community. De Varine and others have also shown that, in some cases heritage can also act as a source of empowerment signalling the independence of a community defining itself at times of crisis and tensions through exhibitions that are statements of contest (ibid. 162-91).[12] Returning to Clearance narratives, undoubtedly the 'idiosyncratic' museums seen earlier can qualify for such a reading through their engagement with contemporary issues, such as Gaelic revival or, more tellingly, land access and legislation at a time when they had not been enshrined in law and the movement for community ownership had not reached its zenith.

More widely, taking Clearance narratives in the museums surveyed as an index, the promotion of cultural tourism has not emasculated the people's memory of the Clearances, based as these narratives often are on the imaginative experience of expulsion, displacement and dispersal. The following chapter will explore the implications of such narratives of the past in identity definition and the reflective role of museums. In particular, the extent to which nuances and inflections in Clearance narratives can be associated with their context of emergence, along with dominant political and cultural issues and discourses.

11 Personal communication, October 2009.

12 See also Davison (2005).

Chapter 8
The Politics of Clearance Heritage

We believe that arts and culture have a central role in shaping a sense of community and civic pride in the New Scotland (Scottish Executive 1999).[1]

The choices made at the MoI to expunge tales of hardships and issues of responsibility from its storyline, the prominence afforded the part played by Scots around the world at the Museum of Scotland which frames its Clearance narrative, or the ways in which folk material culture has been enlisted in various venues over time, underscore the cultural processes at work within the confines of the museum – be they national or local. Museums are key instruments used to communicate the essence of a place and its people, and the sifting of past evidence in the construction of narratives of the Clearances exposes the crucial importance of modern concerns. Museum narratives of the past, play a crucial part in crystallising ideas of origin, cohesion and togetherness, in giving material shape to an 'imagined community' (B. Anderson 1991) by instilling a sense of common experience and shared values within social groups. They are loci of expression of cultural identity, a fluid and protean concept, which can best be grasped if both the focus group of the story and context of retelling are unravelled.

Culture, identity and museums

In 1993, the *Charter for the Arts in Scotland*,[2] in its preamble, provided a succinct definition of culture that concentrated on the centrality of culture in identity formation through the emotions it elicits within individuals and groups; culture was apprehended as a constructive shaping force, enhancing and fortifying positive self-images:

> Culture is the key component in defining human identity at individual, community and national level. Through our culture we seek fulfilment and pleasure and

1 This was the statement of principle and intention preceding the National Cultural Strategy published in 2000. Available at: http://www.scotland.gov.uk/nationalculturalstrategy/docs/csnse-00.htm.

2 The *Charter* presented the conclusions and recommendations for Scotland of a survey of the provision and challenges facing the arts. It was part of a wider investigation commissioned by the Minister for the Arts in 1991 in order to elaborate future strategies and policies for Great Britain. In Scotland the consultation exercise was undertaken and administered separately.

> the sense of being a whole person and a whole people (Scottish Arts Council et al. 1993: 3).

After devolution, when, in 2000, the Scottish Executive delineated its National Cultural Strategy, they chose to endorse UNESCO's definition cited in 1982 in its Declaration on Cultural Policies; wider, more holistic, it tried to pinpoint the components of culture:

> In its widest sense, culture may now be said to be the whole complex of distinctive spiritual, material, intellectual and emotional features that characterise a society or social group. It includes not only the arts and letters, but also modes of life, the fundamental rights of the human being, value systems, traditions and beliefs (Scottish Executive 2000: 4).

The Scottish Executive's position shows the contingency of the conceptualisation of culture as it spoke, at the same time, to a conception of cultural identity laying stress on diversity, appreciation of cultural minorities, pluralism and dialogue enshrined in UNESCO's policy document (1982), and to New Labour's discursive emphasis on social inclusion. The First Minister's foreword presented the Executive's approach as 'inclusive and diverse', while Scotland's culture was strongly articulated upon the impact of migration into and out of the country making it a 'rich, complex blend of the indigenous and the international' (Scottish Executive 2000: Foreword, 5). Where it outlined the contribution of culture, the vision was made capacious enough to accommodate not only culture as a motor for economic regeneration and development, notably through the benefits expected from cultural tourism, but also to serve the Executive's proclaimed goals of social justice and equality (ibid. 2). Scotland's museums were to participate in this reforming agenda and be made more accessible in view of their 'important contribution to education, scholarship, citizenship, social inclusion and tourism' (ibid. 29). Beyond the economic dimension of museums as part of the service industry, the conception of the Scottish Executive corroborates the perception of museums as spaces diffusing ideas and fostering acceptance of, as mentioned here, inclusion and diversity, crucial in the case of Gaelic language and culture; but the reforming agenda, in the context of early twenty-first-century devolved Scotland, could also be easily transposed to the land question and related issues of power relations in the crofting districts, which the Scottish Parliament was about to tackle. Rather than a prognosis on a future programme for museums in Scotland, the National Strategy can be seen as a forceful illustration of the long-standing social and political roles of museums.

In late twentieth-century Britain, as seen in the previous chapter, museums embraced the social discourse of diversity, inclusion and renewal, which were, in devolved Scotland, enhanced and expanded to meet Scotland's new political identity. In the early years of last century, their development and role were often predicated upon the national ideal and the forging of a national *British* community. The pivotal part of national museums acting as a centripetal force in identity construction has been amply demonstrated; Anderson has shown how profoundly political the 'museumizing' process was to colonising powers and post-independent states (B. Anderson 1991: 178-84) and Boylan identified the national museum as one of 'the four most vital symbols of independent nationhood, and most vital instruments for trying to keep people together and create a true nation' (Boylan 1990: 30). It is from this vantage point that the creation of a British folk museum was mooted in the early issues of the *Museums Journal*, which started publication in 1901. By then, the argument surrounding the need to make museums less socially exclusive, less the preserve of the cultured and the scholar, but spaces for 'rational recreations' (Message 2006: 92) open to all had been well practised (Coombes 2004). Underwriting the tripartite functions of museums – 'investigation, instruction and inspiration' – were, at one and the same time, the project of widening the museum's audience to the lay public, in keeping with what was termed the 'democratic spirit' of the age (Bather 1903: 93-94), in short transforming museums into 'places of popular culture' (ibid. 75), and of promoting a sense of cohesive national identity within the British Isles through the representation of the history of culture of the British Islands.[3] These twin aims coalesced in the vision of a National Folk-Museum which was advocated using nationalist discourse:

> ... in most European countries national pride has found expression in the formation of national collections, and a folk-museum is a prominent and patriotic feature in very many of the continental cities and towns ... These folk-museums and departments are devoted to the preservation of objects of strictly national interest, and serve as repositories in which may be seen and studied not only the domestic and other appliances and the various material relics of bygone centuries, but also characteristic features of the more recent culture and social economy of the peasantry, the *backbone of every nation* ... Such national ethnographical collections are a popular success, appreciated all the more for being easily understood by the unititiated and for appealing to local patriotism (Balfour 1909: 8-9, my emphasis).

Up until the 1920s, the Museums Association pressed for its creation resorting to the same rousing rhetoric in which patriotism and a sense of common identity found their source in the representation of a shared past rooted in images of the country. The benefit for the nation, it was felt, would far outweigh the financial cost that such a project inevitably entailed as, 'to increase the love of one's native land [was] no unpractical or purely sentimental aim' (*MJ* 1930: 380). As Raymond Williams commented, the 'country', in English, can be made to stand for the

3 See early issues of the *MJ*, in particular pieces by Henry Balfour (1904, 1909, 1912) and A.S. Bather (1903, 1911, 1916) who both served as President of the Museums Association.

nation – the Country – not only in linguistic terms, but also through the powerful feelings and associations it generates (R. Williams 1973: 1). It was into feelings of common origin that proponents of a national folk museum sought to tap at a time when material remnants of traditional practices and skills could still be found and preserved.

That the project never actually materialised, at least along the line initially sketched out, certainly owed to expenditure compounded by such pressing concerns as financial and international crises. From its inception, questions were not only raised as to its practicality but, also, its cultural rationale: what to make of the epithet 'national'?[4] By 1912, a significant shift was apparent, from a national museum representing the people of the British Islands, to one illustrating 'the culture history, and the modes of life in times past, of the English peoples' (Balfour et al. 1912: 222). When the Royal Commission on National Museums reported in September 1929, it deplored the absence in the British Isles of a folk museum comparable to those established in Scandinavia and Holland and recommended that one should ultimately be erected in each England, Scotland, Wales and Ireland, the 'most urgent [being] that relating to England'[5]; a Folk Museum Committee was set up to consider the establishment of a National Folk Museum in or close to London but its work was interrupted by the financial crisis of 1932 (*MJ* 1931, Mortimer Wheeler 1934: 191). For Cyril Fox (1934: 115), Director of the National Museum of Wales, 'so diversified is our national life, so diversified the expression of it, that as soon as one central museum is safely launched we shall have to meet the demand for Regional Museums'. His projection highlights the major obstacle of a grand national project: the sheer diversity of building techniques and materials within Britain and, within its various components. So aside from the feasibility and expenditure of such a project in inter-war and post-1945 reconstruction years, the common memory of a notional 'folk' was not easy to manufacture. As a 'national costume', the past culture of the folk, despite its rallying patriotic colours, showed uneven seams and conspicuous stitching, not only at a geo-political, but also social level – the rural folk often overshadowing their industrial relatives.

If a National Folk Museum along Scandinavian lines did not – and was not – to see the light of day, folk collections were being, or continued to be, assembled and displayed in larger institutions such as the National Museum of Antiquities in Edinburgh and a folk movement did take root in the UK, notably through local and individual effort. And indeed, such local efforts were continuously recorded and praised in the pages of *The Museums Journal*; among them was I.F. Grant's adventure with *Am Fasgadh* which conveys a sense of her vision, determination and struggle against bureaucratic rules (Grant 1936: 347, *MJ* 1945). But the HFM was a private endeavour, and lacked the seal of official recognition that nationally sponsored projects enjoyed and, in this, it differed from the folk movement as

4 See Charles Madeley's comments in the discussion following the presentation of the 'pavilion system' (Karlin 1904: 394).

5 Royal Commission on National Museums and Galleries (1929: 60).

it developed in Wales. From the thirties, the voices which gradually came to define 'folk culture' as a discipline and the province and nature of folk museums originated from Wales, most notably through the numerous articles authored by Iorwerth Peate. These also trace the contribution of the National Museum of Wales which, in Britain, led the way in creating a Department of Folk Life in 1936, under his keepership, and in pressing for the establishment of a Folk Museum which, from 1946, he developed after St Fagans castle, gardens and grounds were gifted by the Earl of Plymouth to the National Museum of Wales.[6] So in the folk museum movement, Scottish national museums certainly lagged behind. As seen in Chapter 3, Peate's counterpart was only appointed in 1959, and, in addition, legal constraints hampered the powers of the Trustees of the NMoAoS as regards the creation of an open-air museum, in stark contrast with the status of the National Museum of Wales which gave it more scope and freedom for initiative (*MJ* 1946: 106). Further, Peate's vision which married renaissance of the Welsh language with reverence for Welsh rural society and neglected the more urbanised and industrialised English-speaking Wales was able to draw on the values at the heart of Welsh nationalism at the time (Aughey 2001: 115-16, Stephens 2006).

The construction of the rural 'folk', their culture and history, often stem from defensive positions or from pursued social or political ideals, through the prism of which they might be recast. As shaping forces, museums often have to contend with the cumbersome cultural baggage of their diverse audiences, baggage loaded with mixed emotions including ambiguous, let alone, derogatory self-images. At the time of the 1930 Inverness Highland exhibition, when I.F. Grant, the trailblazer in the Scottish folk movement, was 'campaigning' for the establishment of a Highland Folk Museum, she saw it as fulfilling a dual purpose:

> In the first place it would help to give townsfolk from all over the country some flavour of the past life of their own districts, for there is a certain degree of kinship about the old-world rural life of all folk of Britain; but, in the second place, and far more directly, such a folk museum would help preserve the individuality of the distinctive Gaelic civilisation of the Highlands, with its particular message and value in the materialised age (Grant 1930b: 284).

Grant's vision of an essential, nationally-encompassing, folk culture was filtered through anti-modern and romantic spectacles, and frequently reflected in her bemoaning the changes in agricultural methods and standard of housing in the Highlands and Islands (ibid., Grant 1962). In her vision, the past material life of the Gaels is inseparable from a value system and way of life, undervalued in the modern materialistic age. Therefore her museum project was not primarily to promote a national imaginary, but, above all, it was meant to serve and redress the image and perception of the Gaels, their culture and status in Scottish society. It

6 See Peate (1934, 1941, 1947, 1949, 1954); incidentally Peate objected to I.F. Grant's choice of replica building and favoured the use of originals (1949: 68).

was one of rehabilitation and healing through the medium of a museum exhibiting the material possession and way of life of the Highland folk – a way of life, austere setting and primitive implements which connoted simplicity, resourcefulness and, most of all, poverty. While her reminiscences of the creation of *Am Fasgadh* are infused with her own passion and the sense of pride with which, she felt, Gaelic heritage was held by its inheritors, they also resonate with the cultural heritage of the experience of poverty and, far more significantly, its construction and treatment in society. Time and again, she records the reticence of Highland people to see their penury and archaic ways of tilling the land or living put on display (Grant 2007: 17, 28, 60-61), and how this impeded her collecting forays. She also frequently underscores, with irritation, how this affected the condescending views of Lowland visitors to the museum (ibid. 106, 187). This was an inheritance of shame which, it could be argued, had a long pedigree, whether linked to the stigma attached to 'pauperism' in nineteenth-century Scotland – most vivid during the acute famine years of the mid-century – or the view of the crofting districts as over-reliant on state-aid since the introduction and extension of crofting legislation. The schemes of slum clearances pursued by the Ministry of Health also exacerbated this perception. Shame attached to material conditions of living and obstructing museum memorialisation of the past is an argument echoed by others with similar goals. Witness for instance the press article written on the opening of Skye Cottage Museum – as the SMoIL was initially known – and commenting on a similar attempt made on Lewis:

> ... preserving the ethos of a vanished way of life is of increasing importance ... Strangely enough, many people opposed this idea. It seems many native Lewismen want to wipe out all the memory of the island's past history. In their understandable zeal to modernise conditions in the island they seem to have almost a shame of the past (D. MacLeod 1965).

This was a time when 'white houses' were fast replacing their black counterparts in the island landscape. As with I.F. Grant's, the need to preserve the past did not only spring from a sense of urgency due to the accelerating pace of material change, but from a vision imbuing 'traditional' Highland and Island society with norms and beliefs at variance with and even subverted by modern materialistic values collapsed into modernisation and consumerism. To a large extent, readings of Gaelic culture have tended to conform to anthropological binary oppositions such as 'tradition' versus 'modernity', 'solidarity and community spirit' versus 'materialism', terms encapsulating a way of life and ethos deemed all the more precious and ideal as they receded into memory.

I.F. Grant's reading of Highland society and culture was, in addition, premised on the assumption that: 'that society, of all ranks, was of aristocratic origin and not peasant ...', a point which she felt, 'was not sufficiently understood or taken into account in specialised studies of individual aspects of Highland culture or of modern local conditions' (Grant 2007: 17, 148). *Am Fasgadh* was one means, in the cultural sphere, of offsetting a deprecating self-concept by buttressing a less murky social identity. Her displays were meant to convey, to the local public she hoped to attract, particularly the younger generations, an understanding of the close inter-dependence of the culture, society and environment of their forebears. For instance, she drew a clear link between what is now described as tangible and intangible heritage in order to celebrate the strength and creativity of the Gaels. Contrasting their rich musical and poetic heritage to their primitive agricultural implements and dwellings set against a challenging environment, she concluded that:

> It is the glory and significance of the mental and spiritual achievements of the ancient Gael that they were fashioned in dire distress by sweat and blood – the sublimation of a struggling people (Grant 1930b: 284).

Looking into her reading of Highland culture throws up issues, revealing of attitudes as regards the Highland region, which impinged on interpretive and memorialising endeavours. This provides useful pointers in order to understand the way in which the Highland past and narratives of the Clearances have been constructed in museums, both local and national.

Interpretive communities and Clearance narratives

Museums, and more widely the cultural sphere, are being defined as dynamic forces fostering new social identities. Additionally, as seen in the previous chapter, the discourse of community which came to legitimise the retreat of the state and promotion of self-help initiatives from the eighties was absorbed by cultural actors and organisations. Community became a core criterion for a variety of agencies to which museum-type projects could appeal for funding, be they at national or European levels. Such projects had to show that they benefited their communities and had to make a case for their contribution to local regeneration, input and ultimately empowerment – which became a conspicuous buzz word.

As already seen Clearance narratives are modulated according to the museum community they serve along with the audience they seek to address. But as a concept community can be as narrow or spacious as identity. Like identity it is multi-layered, open and dynamic (S. Watson 2007b). Museums are places where the paths of different communities intersect and where they often interact (Davis 1999: 24-44). A small museum in a remote area may more easily capture and project the identity of its locality and surrounding society than a national institution, yet it is frequently placed in a zone attractive to tourists and is a port of call for an international audience. Not only do Highland and Island museums owe their continued existence to their being positioned along tourist routes but, for genealogical venues, their *raison d'être* is precisely dialogue and interface between different individuals and groups, assembled in the case of clan museums under the banner of tribalism. In 1994, T.C. Smout (1994: 107) observed that 'modern Scottish identity is much more firmly allied to a sense of place than a sense of tribe', in contrast with the way Scots in other parts of the world, notably in America, constructed their Scottishness. Over 15 years later, it seems that this sense of tribalism is being thoroughly capitalised on in the heritage industry, and endorsed by the then Scottish Executive, now Scottish Government, in their decision to mark the 250th anniversary of Burns' birth in 2009 as 'Scotland's Year of Homecoming'.

One of the most useful concepts to examine the way Clearance narratives are deployed in different museums is that of 'interpretive communities' defined as:

> ... located in relation to interpretive acts. It is only through the common repertoires and strategies used in interpretation that such communities can be recognised ...
>
> Interpretive communities are fluid and unstable. Their membership changes as individuals revise their affiliations and redirect their interests (Hooper-Greenhill 2000: 121).

It is possible to identify a variety of interpretive communities impacting not only on Clearance narratives, but also on their material basis, such as the collections accumulated over time whose initial purpose might not tone in with the interpretive community spirit at work in present-day Scotland. Two broad categories seem to stand out: communities brought together by a shared sense of history and experience (past or present) and museological/collecting communities. Nuances in the Clearance narratives found in such distinct venues as the MoS and the NMoRL can, in part, be attributed to the communities they serve and the hue of their time and context – the issues that undergird them reflecting present concerns.

The MoS, opened on St Andrew's Day 1998 shortly after devolution had been endorsed by the Scottish electorate, and conceived at a time of intense national debate about the status and future of Scotland within the UK, could not fail to be read in terms of its contribution to the definition of nationhood and Scottish national identity.[7] To what extent was the approach adopted in the treatment of the Clearances at the MoS – fragmentary, mainly restricted to emigration and to famed events – affected by the context of its emergence? In spotlighting the routes followed by Scottish emigrants and a world map studded with Scottish place-names, in the process, the MoS foregrounds Scotland as a successful and enterprising colonial nation (dimly part of the British sphere of influence), and also as one open to trans-national exchange in the context of post-colonial Britain. However, if trans-national cultural routes are given pride of place, the museum, by contrast, tends to underplay sub-national cultures and traditions, as the examination

7 See the reactions vented in the media quoted in Jones (2000: 10-12) and *Guardian* (1998); on the MoS and identity see McLean and Cooke (2003).

of the section 'Living on the Land' in previous chapters highlighted, foregrounding continuity of tradition subsumed in seamless rural experience. Further, the cultural specificity of Gaelic Scotland – so often co-opted in the racialising discourse of some social reformers, or clearing and emigratory policies in the nineteenth century (Devine 1988: 125-26, Gouriévidis 2001, 2003-2004) – seems understated, barely suggested through emigration adverts written in both Gaelic and English and a Gaelic bible belonging to an emigrant to North America. Yet, the opening of the museum occurred at a time when Gaelic was enjoying a new lease of life sustained by a variety of long-standing projects and when, as mentioned previously, the EU, with the Maastricht Treaty, was giving increasing recognition to infra-national territories, however what dominates the representation of the rural transformation of modern Scotland is a tale of shared experience rather than cultural diversity and wealth.

To elucidate the museum's narrative choices, it might be useful to return to accounts of its conception and maturation; they underline that the objective of the new museum was to showcase Scotland's culture and history through the national collections which had been accumulated over time: 'the centrality of the object [being] the guiding principle' (J. Calder 2000: 42). At the same time the NMS were pressed to 'answer public expectations by covering, for example, the Clearances ...' (McKean 2000: 102-103). These expectations already seemed patent when, much earlier, I.F. Grant (2007: 55) was taking visitors round her collections, as she 'was often struck by the strong feelings that 'The Clearances' arouse even in people otherwise not much interested in Highland social history.' They are confirmed by McCrone et al.'s survey (1995: 167-68) delineating the characteristics of Scottish heritage in which the specific heritage of Highland clans and the Highland Clearances were recurring 'non-physical' features. The resulting cursory representation of the clearing process at the MoS seemingly might be resulting from a re-formulation of objectives; all the same, ultimately, it gives the period little significance within the national narrative of social and economic transformation and much within the construction of Scotland as a diasporic homeland – emigration becoming its most memorable legacy. Besides limitations imposed by space, it is a posture which appears to have much to do with the economic potential of global tourism, particularly genealogical tourism, and the funding sources of the MoS project, rather than the availability of artefacts or evidence.

As seen in Chapter 4, at both the MoS and the NMoRL, ideas of dispersal, uprooting and loss of familiar and affective landmarks take pride of place – the people's experience of the Clearances. However, whilst at the MoS the process serves a global, diasporic reading, at the NMoRL, Clearance as a notion is recycled to serve a larger national constituency as changes in Lowland rural life were recast within this interpretive repertoire, when the Clearances section – initially purely devoted to the Highlands and Islands – was revamped in 2006. The themes of displacement, fragmentation of community groups and 'emptying land' provide the unifying strand, with the discourse of Improvement or rationale behind such social dislocation, discredited as a means of emplotting the transformation of rural society and landscape and relegated to a backdrop. In late-twentieth and early twenty-first-century Scotland, the Clearances have a profound legacy and are embedded in Scottish culture as a wound with strong national resonance. As a marker initially confined to the Highlands, now applied to the Lowlands, 'the Clearances' might be another instance of the colonisation of Scotland as a whole by a powerful Highland sign.[8] On the one hand, this reinterpretation certainly casts light on processes and experiences deemed overlooked or little emphasised in other parts of Scotland and places issues of power and social engineering squarely into the story of Scottish industrial and agricultural 'revolution'. On the other, unfortunately, it obscures major agricultural, legal and, ultimately, socio-cultural differences such as the long-lasting mental repercussions of the clan system on the peasantry, the organisation of Highland resistance and, more remarkably, the specificity of crofting and the separate body of legislation regulating it since 1886, central to modern-day Highland discourse and culture. More tellingly perhaps, such a narrative of national fusion screens out the cultural fissures which saturated contemporary discourse such as, at one end of the spectrum, the racial construction of the Highlanders as inferior and, at the other end, the perception of Lowlanders – employed as emblematic 'shepherds' on Highland estates – as willing and active instruments of the clearing process. Socio-cultural fault lines long characteristic of Scotland's cultural, but also ethnological, discourses tend to be obliterated.

Both the MoS and the NMoRL depart from the Enlightenment view of agricultural transformation in their Clearance narratives, a view which underscored improved food production alongside scientific and material progress. Interestingly the museums' collections of rural artefacts partly stemmed from the recording and exhibiting initiatives of organisations actively engaged in amassing objects and data for the dissemination of those ideas. Museums are, in fact, loci where various museological or collecting communities come into contact through the re-ordering and re-contextualising of their material over time. Some of the collections of rural artefacts at the Mos and the NMoRL were inherited from the National Museum of Antiquities of Scotland, in particular the models and machines assembled by the Highland and Agricultural Society of Scotland set up in 1784 (Fenton 1962). These examples were meant to illustrate technological innovation, celebrate inventors and their creations and, above all, encourage similar pursuits through, for instance, the award of prizes; they were three-dimensional records of Scotland's achievements, showcasing progress and advance and, because of their educational and publicising dimension, spoke to the present and the future, rather than the past. If such sanguine narratives of improvement and progress are strongly featured in both museums – the Mos dedicates a part of its Enlightenment section to the Society's work, and at the NMoRL, the visitor is led into the museum's exhibition through a remarkable collection of the Society's models – neither museum chose

8 The idea is borrowed from McCrone et al. (1995: 201).

to lodge the Clearances within these themes, thus offering fragmented (NMoRL) or lopsided (MoS) accounts of events.

Also inherited from the National Museum of Antiquities was the collection accumulated under the incentive of Sir Henry Mitchell from the mid-nineteenth century, the backbone of which came from the Highlands and Islands (Fenton 1962, Cheape 1979). The thinking behind its assemblage was conditioned by Mitchell's methodology which was shared by other European scholars at the time: looking for marks of pre-industrial ways still in existence in the present in order to shed light on earlier times.[9] Some artefacts had also been bequeathed as curiosities or oddities snatched up by travellers to the region. Underpinning these collections were contemporary evolutionary concepts derived from the Darwinian debate which, allied with the Enlightenment notion of progress, cast on highland and island exhibits an ethnographic gaze not so different from that moulded overseas by colonial collectors intrigued by the 'primitive' and 'exotic'. By and large, neither museum attempts to critically expose the history of their collections and their influence on cultural constructions, at least with respect to 'rudimentary' agricultural implements. Yet, a substantial part of the interpretation of Highland culture at the MoS hinges on the deconstruction of images and symbols, in particular those prominent in the catalogue of Scottish mythologies. The late twentieth-century museological community took their cue from the literature and exhibitions spawned by the 'Invention of Tradition' argument, notably 'Scotch Myths', a satirical exploration of stereotypically kitsch images of Scotland. Originally on display at the Crawford Arts Centre in St Andrews in 1981, it took Edinburgh by storm during the Festival and was later turned into a documentary film. Not only are the Clearances introduced from this premise, as part of the debunking of the myth of romantic wilderness, whose underside is described as the 'mass displacement of people in favour of sheep and later deer', but also another part of the exhibition, devoted to the dialectics 'images and realities', displays Queen Victoria's contribution to the process. Whether both national museums are demythologising long-rooted romantic images or shunning the Improvers' frequently complacent vision of agricultural change as inseparable from progress, importantly relations of power and landlordism underpin their narratives of the Clearances, however cursory and fragmented their treatments of the period might be.

Small museums with more restricted locational frames of reference, also often strive for cohesiveness and coherence through the stories they tell. The Miles Report noted that very small museums 'were born from a sense of community' (Museums and Galleries Commission 1986: 9), although small communities, as Davis suggests, may also be prey to social tensions and fractures which a museum project may well (re)kindle (Davis 1999: 31). Yet, with the increase of people choosing to move into the region – dubbed 'white settlers' by some – Highland

9 His views were developed in his Rhind lectures delivered the 1870s, later published in book form (Mitchell 1880).

and Island museums have also been apprehended as crucial social tools, a means of countervailing the cultural dilution generated by incomers settling in remote communities.

> Museums have a fundamental role in community development (…) With a large body of "incomers" the pressure to change common values can create stress. However, by giving the community a chance to overtly express their values in a neutral environment, and by giving the incomer access to the underpinning history that has led to the development of these values, a sense of pride and belonging can be engendered (G. Watson 1995: 258).[10]

Intertwined with a sense of the past is a sense of the core values of a place, and museums are regarded as holders and transmitters of a community's memory and identity defined through the narratives constructed.

One of the questions put to the staff running the museums surveyed included commenting on the importance of the Clearances period to Highland and Island identity (Appendix section 11). While there was unanimous agreement as to its importance, what transpired amongst those that expanded, was the unstable, group- and context-bound nature of identity and the resultant contingent and fluctuating significance of the past. Identity variables included, on the one hand, constitutional developments such as the re-establishment of a Scottish Parliament, and by extension relationships with England, on the other, social parameters, for example the input from 'overseas' Scots as well as 'incomers' to the region.

> *Glengarry HC*: A sense of Highland, and indeed, Scottish identity would seem to be strongest amongst emigrant communities. The period post-clearance/emigration saw an evolution of a somewhat romanticised, unbalanced view of the times and event amongst non-emigrants who, even today, may sometimes see themselves as disadvantaged heritors of a noble tradition and culture.
>
> There has been a dissipation or dilution of culture in some areas, in part through the influx of people from other backgrounds. However, recent political devolution has possibly stimulated a resurgence of a sense of "Scottishness" in today's population.
>
> *Strathnaver Museum*: Paramount. It is of crucial importance to the diaspora which has left a skeleton population in a once populous area. Contrast the over-populated central belt which is unique in most small North Sea countries.
>
> *Mallaig HC*: This is difficult to define – most people see the Clearances as a purely 'Highland' phenomenon, unaware that similar events occurred in Southern

---

10 Graham Watson was manager of the Heritage Section of Ross and Cromarty District Council.

Scotland and elsewhere. Similarly, most parts of the Highlands saw some form of Clearance, but most people are only aware of the "headline" events.

> *Ullapool H&VC*: Too important – in the sense that it is often used by people who want to justify xenophobia and chips on shoulder. Easily misinterpreted.
>
> But on the other hand I think the make-up of the Highlands and better communications are changing so much that outside museums and heritage centres it doesn't have such importance. In other words, it is more heavily a part of the image than identity. I think people who consider themselves Highland live increasingly in today's world and a more cosmopolitan one, but the image put across to tourists is still one that relies on romanticising the period and its hardships.

In this last answer hinting at the manufacturing of the past for public consumption, the idea of indulging visitors' expectations with a deliberate 'romanticisation' of *tragic* events is also voiced, drawing attention to the shifting relevance of the past in identity production, particularly in view of its significance for the tourist industry. Also percolating through this answer is the importance of dispelling distorting preconceptions – 'xenophobia and chips on shoulder' – a view shared by the respondent at the NMoRL who mentioned that the Clearances were, in some measure, 'a well-known part of Scottish History, though often misunderstood ... and can [mistakenly] be seen as the "English" clearing the Scots off their own land'.[11] One final element was underlined by a respondent who mentioned the extent to which the Clearances had contributed to determine attitudes to the land and had moulded local mind-sets providing the inhabitants with a stock of experiences from which they drew and positioned themselves.

> *Arran HM*: There is definitely a very strong emotional connection with the injustices of feudal landownership and the clearances epitomise this and are rooted now in the people's psyche.

Clearance narratives and the politics of the land

If, under the influence of the 'New Museology' movement, museums have been reappraised and come to be seen as loci for debate and interaction, they also remain legitimising institutions, particularly as regards the versions of the past they sanction and project to a varied audience. As the respondent for Arran HM observed, as a subject-matter the Clearances are laced with issues of landed power, of the morality of landed policies, eventually resting on the nature and source of national land legislation. That they were recently incorporated into the nation's

11 Private correspondence, 27 February 2009.

narrative – albeit very superficially and with different emphases – in the national museums surveyed is museologically and historically meaningful. The same applies to some of the small venues surveyed, some of which, privately owned and without any affiliation with official agencies, also mirroring a sense of the debates and concerns which have affected the social and political landscape of the region in recent decades.

Two of the questions asked in the survey were connected with the perception of the Clearances in local society and culture (Appendix section 11). Answers to the question inquiring as to whether exhibitions showed the Clearances as having shaped the present-day Highlands and Islands focused on the demographics of the period in terms of population distribution and decline (Arran HM, Mallaig HC and Glengarry HC), and on their physical impact with respect to landscape, land use or town planning and development (Arran HM, Glengarry HC and Timespan). To some extent, this confirms a tacit vision of museum exhibitions as essentially apolitical. It is in some of the answers to the question concerning the importance of the period in Highland/Island history and culture that the cultural and mental impact of the Clearances came to light. All the respondents agreed on the significance of the period in regional history and culture but emphasised different elements; again demographic effects (MoI and Mallaig HC) were cited along with socio-economic factors (Glengarry HC), such as the Clearances marking the ultimate breakdown of the 'feudal system'. But much wider ramifications were also brought forward:

> *Strathnaver Museum*: This was a formative period and is responsible for most of the way of life within the Highlands today.
>
> *Timespan*: Major importance as it informs social and political attitudes today.
>
> *Seallam! VC*: it has to be important in that it extends the history of Highland & Island people beyond their own area.
>
> *Ullapool M&VC*: It is important since it reflects both a shoddy attitude and treatment of the Highlands that existed for a long time, and the first real breakthrough in righting that. It also has a strong influence over the culture. However, there is often a danger that it obscures other aspects of history and culture and is presented lopsidedly.

Whilst the flow of history has turned the Clearances into a mandatory feature in local displays, they are also seemingly perceived as culturally loaded by some museum staff as alluded to in the last answer. For Ullapool H&VC, the Clearances in fact were not especially singled out as having shaped the present-day local area, at least 'no more than anything else'. However, in the majority of venues, the Clearances as a 'formative period', signal a new way of life either at home or abroad with exhibitions often portraying the radical transformation of the physical and social landscape of the region: local society and identity being reshaped. As such the Clearances are frequently cast as foundational. They are also politically symbolic in their association with landed power and have long been appropriated as an emblem of inequality and oppression in political discourse.

Engaging with the Clearances, in many museums, therefore also meant addressing issues of power, in particular landowning rights, regulation and status linked with national legislation. In the late twentieth century this also chimed in with developments in crofting districts such as the movement for community ownership of land and calls for land reform, together with the shape future measures were to take in a devolved Scotland. The 'strong emotional connection with the injustices of feudal landownership' embedded in local mind-sets which the respondent for Arran HM underscored, brings to light the power of such a sensitive and contested issue as the Clearances in buttressing what could be best described as a sense of grievance about social injustices, with its echoes resonating well into the present. Long-term occupation of the land called to a halt by Clearance pervades museum narratives and trails. Furthermore, as delineated in survey answers, in most venues, the outcome of the Clearances is addressed from a physical and demographic perspective with recurrent allusions to the making of the region's crofting landscape, but in some museums, parallels or connections are drawn with present concerns, in particular crofting issues.

Referring back to the narrative strands outlined in Chapter 4, this perspective is mostly found amongst exhibitions developing the local strand of Clearance narratives with specific emphasis on crofting and the Land Wars: *Aros*, Ullapool M&VC, SMoIL, Colbost, *Seallam!* VC. They are museums whose narratives take the visitors on a journey through time ending with an overview of modern days, the Clearances period being related with socio-cultural issues which are charted to the present. Most striking among them are the 'idiosyncratic' museums where a variety of late twentieth-century issues are debated and which bear the imprint of their authors. *Aros* was another instance; its clear agenda was to 'celebrat[e] the ordinary people of Skye. Generations of men and women, who endured the misery of past centuries, so that the "Misty Isle" could flourish today' (*Aros* n.d.). In its former exhibition, it was not the birth of crofting – largely resulting from landowners' reforming policies – which was celebrated but its survival to the present owing to crofters' action and gains following the land war; it was the resilience and struggles of past generations of crofters which were commemorated.

In all the venues listed above, the protection of crofting as a culturally-specific and valuable way of life is foregrounded with the Napier Commission and subsequent legislation hailed as crucially important in the history of the area. At SMoIL the decline of Gaelic in the nineteenth century is connected to the expulsive forces of the Clearances and its renaissance to the renewed economic vitality of the crofting districts and government support to efforts at linguistic development; similarly the birth of crofting is tracked back to the Clearances period and its distinctiveness seen as its greatest asset worth the protection and assistance it slowly came to enjoy over time – a possible model for the survival of small farmers elsewhere in the world:

> Crofting of course, is more than a system of land tenure and more than a means of making a living. It is a unique and satisfactory pattern of life in a part of the world that is still refreshingly different.

The vocabulary selected clearly underscores the singularity of crofting and by extension that of the Highlands and Islands, thereby warranting differential treatment with regards to the land question. Interestingly, Colbost and SMoIL anchor their story in the local and the particular and speak volumes to the distinctiveness of their social and political context. Both estates were purchased by the Congested District Board in 1904; the tenants of Glendale however chose to acquire their land whilst the Kilmuir crofters declined to do so (*The Clarion of Skye* 1955). In the potted history on display at SMoIL, the tone appears somewhat defensive when the conditions of twentieth-century crofters are depicted and their legal situation justified:

> The crofter is not more heavily subsidised than the industrial worker of the same economic status in a Council house. Indeed it costs the State little more than half as much to house a crofter as it does to provide local authority house.

Changes in housing policy in the crofting areas – extension of loans and recipients – was an issue which was repeatedly pursued by the Crofters Commission in the early 70s, incidentally eventually sanctioned by HIDB (MacCuish 1976-1978: 316-17). The museum's owner was a member of the Commission at that period[12] and the text bears out a personal and institutional orientation, at a time when more wider reform of the crofting legislation was also being envisaged. At Colbost, where late twentieth-century local land issues such as the sale of the Cuillins, are juxtaposed with late nineteenth-century agitation and subsequent legislation affording protection to crofters, the contextual notice introducing the exhibition, when last visited in 2007, still paid tribute to Glendale's unique achievement as 'the only crofting community in full ownership of its land' – a particularity now outstripped by the late twentieth-century movement of community ownership and enshrined in the 2003 Land Reform Act. Unusual when first opened, by the turn of the twenty-first century, such a view had become the credo of the quango HIE (*Annual Report 2000-01*: iii) and Scottish political establishment. These small folk museums, voicing the personal views and experiences of their owners, also stand as repositories of the debates and issues which surrounded late twentieth-century land reform. In those cases, museums explicitly acted as public platforms relaying alternative views on the part of local cultural actors and contributed to public discussion. It was also a heritage crafted and located outside geographical and institutional centres.

As McCrone (1997) observed 'landownership' was long seen as 'a problem of Highland estate well away from where most Scots live', and until recently, its most

12 Personal correspondence, 16 December 2008.

potent historical symbol, the Clearances, was not featured in national museums. In local heritage, serious and relatively extensive treatment of the period only emerged in the late seventies, particularly with Strathnaver Museum and the nearby trail round Rosal township, situated where some of the most notorious events took place. The sense that this part of the Highland past needed reclaiming and asserting was shared by many individuals and groups who did not necessarily overplay its political potential, but certainly presented an alternative story to that of stately homes, sporting estates or un-contextualised folklife.

One of the most active – in both cultural and political senses – of these individuals was Rob Gibson, who, as part of a group named Highland Heritage, tried to encourage visitors to the Highlands to travel along a Highland Clearances Trail, following routes outlined in a leaflet first devised in 1981, later expanded and updated. In it, travellers found grid references to clearances landmarks: the ruins of pre-clearance homes and other archaeological remains, buildings associated with specific episodes, monuments dedicated to the period or people related to the events – victims and perpetrators of evictions – and museums. It disclosed the cultural landscape of the region so-called wilderness and told a story of recent human occupation, displacement, resistance and population drain. The guide was described in the press as inviting people to consider the 'seamy side of local history' (*WHFP* 1981), but it seems that the STB was initially reluctant to distribute it because of its political and partisan nature (Gibson 2006: 21).[13] An introduction gave the visitors 'clues' to understand today's Highlands:

> The clusters of ruined homes which litter the Highlands and Islands are an eloquent reminder of the people's experience at the hands of the landlords. Improvements by the lairds were for their profit so inevitably most of the people became surplus to the requirements of the sheep and deer forest economy. Even when the townships were deprived of their lands to such an extent that people found it in their interests to leave and so clearances were effected without exciting much public attention or without causing any great outcry, the weight of evidence provides lessons for us today. Whether in writing, in song or in the memories of the dispossessed these lessons of the Clearances relate directly to the condition of the battered little nation which Scotland is today. The removal of so many Highlanders from their homeland was not only the destruction of Gaelic Scotland but as Neil Ascherson has put it "one of the grand crimes of modern Europe" …
>
> On behalf of the Highland Heritage I hope this publication helps to bridge a noticeable gap in the popular understanding of Scottish local history (Gibson n.d. 1).

---

13 I purchased the 3rd edition of the leaflet from an STB outlet in the Highlands in 1989.

The guide's narrative was an index to the growing strength and cultural potency of the people's vision of the past found for instance in historical publications since the sixties – footnoting many of the leaflet's locations – but also part of the much longer battle waged by Gibson to see this side of history disseminated and remembered. Flagging a range of clearances sites dotted on the region's map was a symbolic gesture of repossession, repossession of the land through its past local history and more recent sites of memory; it was also a means of using heritage to call attention to wider national issues. Gibson's involvement in nationalist politics and campaign in favour of community control of their own affairs (Gibson 1978) was complemented by high profile heritage-related actions, notably through the controversial campaign to have the Statue of the Duke of Sutherland – an emblematic figure of the Improvement spirit – removed from the top of Ben Bhraggie near Golspie (Gibson 1996, Tyrrell 2003).

In the final analysis, as seen in Chapter 2, the Clearances have been a powerful sign used in political discourse calling for land reform and an end to Scotland's 'feudal' land laws. In 1997 McCrone in his John McEwen's Memorial lecture stated that: 'land, democracy and culture in Scotland are intimately related [and] land reform is actually about a more general process of democratising Scotland'. At the heart of issues of land management and use in the Highlands and Islands were questions of lack of understanding of the needs of crofters along with their lack of control in the decision-making process. In late twentieth-century Scottish political discourse, the national reform of the land system in the future devolved Scotland was construed in terms of greater democratic accountability and consensus, the same values underpinning arguments in favour of devolution (Scottish Office 1997). Land reform became one of the flagship policies of the new devolved assembly and a telling symbol of new statecraft and policy-making for twenty-first-century Scotland. In the words of Donald Dewar (1998), who was elected First Minister of the new Scottish Parliament in May 1999, the legislation to come was to be built on 'widespread public consent' and be part of a 'consensus-building process'; its scope was not to be restricted to 'the highly publicised circumstances of the big Highland estates, but with land-related problems in all their diversity throughout Scotland.' If modern historiography has promoted a re-reading of the Lowland experience of past agricultural changes and the re-discovery of its own clearing process, it also seems that the changing political climate surrounding the future of a devolved Scotland has been propitious to the development of a unifying Clearance narrative. The contrast between expulsive forces and the painful experience of displacement and resettlement epitomised by 'clearance' spoke to the reformist agenda of twenty-first-century Scotland.

Chapter 9
Conclusion:
The Dynamics of Clearance Heritage

Heritage and the collective remembrance of the Clearances

If the legacy of the Clearances is writ large in the landscape and population distribution of the Highlands and Islands and, not least in the label 'crofting districts', in Scottish cultural heritage, the Clearances are inscribed as a national trauma, their resonance extending far beyond their regional confines owing to their continuous multi-faceted reconstruction or re-enactment over time and, in social and geographical terms, to the waves of expatriates that clearing policies produced. As illustrated in Chapter 2, their collective memory, embedded in oral tradition and given official recognition by the Napier Commission, has been strengthened by artistic creations, not least literary narratives (among which modern poetry in both Gaelic and English which space has not allowed to treat in this study), and by their use as a rallying and symbolic tool in political discourse. From the nineteenth century, the Clearances became politicised, notably used by Marx, and subsequently recurrently inserted in discourses attacking property and landownership, striking a chord in both urban and rural Scotland. Gradually their memory has become a collective and somewhat festering wound – frequently reopened with the scalpel of contest and controversy.

In the production of this collective imaginary, historians have been relative late-comers, acting as 'sanctifiers' or 'disrupters' of memory, as their work has largely been dichotomised. In the recycling of the past, the post-memory of the Clearances – transmitted from generation to generation and through local lore – is no insignificant feature. An historian's own personal lineage traced back to cleared tenants or districts may be highlighted to account for his/her interest or own take on the events and policy, tending to show that history-writing can be not only a social but also an intensely personal act.[1]

Heritage work can also imply emotional engagement. Memory, identity and heritage – narrowed to sites where conservation, preservation and interpretation take place – are close relatives and their association has been increasingly investigated. They are closely bound up notions and processes. Of memory and identity, Nora (1997b: 17) says that they are analogous yet different notions. Both are equally fluid and adaptable, historically and socially conditioned. Collective

1 See for instance the revised edition of Hunter's *The Making of the Crofting Community* (2000: 1-2) and James Hunter in McGeachy (2005: 8).

memory is the bedrock of identity and is an essential component in its shaping and construction, easily revealed in the marked concerns of nation-states to unite centrifugal memories under a common banner – be it manifested in the guise of universalising narratives of assimilation or the more recent trope of shared citizenry inclusive of pluralism and diversity – or the activities of sectional groups wishing to affirm their particularities and see them recognised and even endorsed by central power. Within the framework of identity formation, heritage can also clearly be enlisted to embody a common identity and memory (Le Goff 1998). Le Goff (ibid. 13) defines heritage as a process located in time; in Smith's words (2006: 66) it is:

> … not a passive subject of management and conservation or tourist visitation – but rather an active process engaged with the construction and negotiation of meaning through remembering.

Furthermore, she observed how important it is to take 'ideas of memory and remembering' into account when attempting to define heritage as a cultural process so that 'a more nuanced understanding of [its] emotional quality and power' (ibid. 303) might be revealed. Heritage, is not only associated with the memory of a common past, but can also help forge a sense of belonging through the interpretation of past events in which the present finds its root.

Throughout the book's argument, the value and use of heritage, in the promotion or rather re-building of a sense of Highland and Island identity has transpired. In the late twentieth and early twenty-first centuries, heritage, and museums in particular, have been regarded as acting as a social cement for remote areas and groups affected by the influx of newcomers; they have been a means of encouraging a sense of cohesion through the understanding of past experiences, ultimately seen as reflecting present mind-sets. Similarly, the use of heritage and more specifically museum exhibitions in rehabilitating an identity long perceived as devitalised, because it was disparaged at worst, or distorted at best, has been a long-running theme and motivation for heritage practitioners and politicians, from I.F. Grant's early undertaking to the more recent designation of 2007 as Scotland's Year of Highland Culture by the then Scottish Executive. The celebration was both a recognition and an endorsement of the cultural sphere in processes of social and economic regeneration – now an integral part of political discourse internationally.[2]

That the shape and spirit of exhibitions as well as the nature of museum-type venues have, at the same time, impacted on and been affected by a wide array of policies, cannot be doubted. The supplanting of material culture by other strategies of communication, the growth of genealogical tourism and the concomitant

---

2 See J. Hunter's introductory essay to the book of the exhibition (2006). The exhibition, funded by the Scottish Executive, was specially created for Highland 2007 and toured around Scotland.

increase of its significance in regional economic and cultural policies, together with the pressure put on exhibition spaces and museums to respond to community needs and aspirations and be integrated within multi-functional projects testify to museum venues, and more broadly heritage, as part of an active process implying negotiation and adjustment. Timespan, whose redevelopment took five years, presents an interesting example of the conceptual shift that has affected many local venues responding to new funding criteria. Its subtitle is 'a meeting place between our past and our future', and certainly 'meeting place' seems to encapsulate its new social function, the adventure of its revamping described as: '... in every sense a community project. From the moment the people of Helmsdale were consulted, their unstinting support has seen us through this very exciting but challenging process' (Timespan 2008: 1). The venue has been transformed to provide facilities including a community archive and workshop enabling greater local participation in heritage projects, notably through oral history and intergenerational activities (ibid.). In design and concept, the exhibition area responds to the twin aims of flexibility and adaptability, a large part of the displays no longer built around Tussaud-like reconstructed scenes, but panels facilitating change and updating of information and content. In reply to the nostalgic comments of visitors – and B&B owners – deploring loss of 'couthiness', 'warmth and individuality' (*NT* 2008), the organisers of the display emphasised the static and permanent nature of the previous one (Woodley 2008). What the transformation of Timespan highlights is not only the significance of economic concerns such as the necessity to cater for repeat visitors and their interests, but also the contingent and dynamic nature of heritage and its loci, constantly adjusted to new needs and conditions. On a wider scale, Kylie Message, in her exploration of the dealings of the 'new museum' with the discourses of cultural diversity and democracy sees it as 'being reconfigured through attempts to create new "museums as cultural centres", which embody postcolonial frameworks and that seek to manifest a more empirically connected set of relationships with communities' (Message 2006: 201). 'Communities' with its plural form seems particularly apposite when adapted to the Scottish context and Clearance narratives.

From local to trans-national memories of the Clearances

Whilst 'community' has become a central notion peppering modern political discourse concerned with social inclusion and the revitalising of specific areas, it is also a crucial part of the socio-cultural definition of the Highlands and Islands as it emphasises their particularities within Scotland as the locale where the 'crofting community' was formed. When considering the memorialisation of the Clearances in museum spaces, 'community' as a notion becomes all the more significant as it enables to account for many of the interpretive processes which have impinged on the representation of the Clearances. To begin with, initially local 'sociations', – in this instance people brought together by the desire to preserve aspects of their past – were for a long time the wellspring of displays related to the Clearances and helped texture the narratives produced. For some, such as that involved in the creation of Strathnaver Museum, it was the explicit wish to tell a story felt to have been ignored or misrepresented in 'mainstream history' – the museum reflecting an interpretive community sharing a common vision of their own past. Gradually, it seems that the economic potential of including narratives of the Clearances became an attractive proposition for many heritage practitioners, partly responding to the needs (perceived or recognised) of the Highland and Island diasporic community or, in the case of clan museums, of a 'tribal' community.

Further, the legacy of the Clearances lurked beneath the surface of discussions linked with the movement of community ownership and the programme of Land Reform of the late nineties. In the debates surrounding Land Reform, when the past was conjured up, clearing policies and their premise were convicted at the bar of modern democratic concerns. It is worth noting that Timespan, in its revamped exhibition opened in 2008 and in its equally reconditioned storyline, now includes a segment dedicated to the Land League – its leading lights in the region, achievements and present-day ramifications. This part of the display, set in a section considering the Sutherland clearances and their modern implications, no longer relies on stage-craft or curatorial irony; the improving family's lofty and ambitious aims are described and juxtaposed with the evictees' perceptions of removals and resettlement.

Figure 9.1 Timespan, Helmsdale, Sutherland
Source: Laurence Gouriévidis 2008.

Importantly, it shows the past as it reverberates in the present – as indeed does the rest of the exhibition – and concludes on the 2003 Land Reform Act said to be heralding 'an important change in the balance of power between the landowner and crofter'. National developments and the local context are closely entwined; the Gartymore cairn erected just outside Helmsdale in 1981, a century after the creation of the local branch of the Land League, and commemorating the actions of local activists is also featured, as are the words of their leader, Joseph Macleod, condemning a landowning system benefiting the 'interests of the few'. It seems that the New Scotland, as it took shape with first the prospect and then the implementation of Devolution, framed by the ideals of greater fairness and accountability to the people, provided a congenial backdrop to the 'up-dating' of Clearance narratives in local museums featuring their resonance in modern-day Scotland. It was also congenial to the inscription of the Clearances within a national narrative encompassing both Lowland and Highland – as illustrated by the NMoRL – the democratic ideals pursued welding together Scotland as a 'national community' with a shared fund of past experiences.

Narratives of the Clearances articulated in Highland and Island museums along with the trails supporting a couple of them, by and large attempt to project the people's experience of the process. They tell of displacement with loss of familiar environmental and geographical landmarks, of the breakdown of social structure and networks of affinities, and, ultimately, of the tearing apart of cultural mind-maps inherited from a past order in which a sense of belonging was anchored. The subsequent struggle to adapt and build on alternative largely class-based values, often collapsed into the all-encompassing label 'crofting community', is also a patent feature of a number of local museums, notably some of those situated on Skye. Therefore narrating the people's experience has undeniably offered significant ideological or at least probing potential, as regards past and, to some degree, modern issues of political economy, their bases and eventually far-reaching consequences, provided interpretations have also invited questions and thinking, and have not been anaesthetised by the anodyne of condescending, or worse, infantilising drama.

Additionally, the varying inflections and modulations, emphases and omissions contained in Clearance narratives show museums as responsive 'contact zones', placed as they are at the junction of local, national and trans-national communities whose memories rub off on each other, as the MoS strikingly exhibits. Certainly, the people's experience, narrowed down to that of emigrants, seems the natural and logical interpretive repertoire for clan museums whose purpose is to act as focal points for their geographically-dispersed members. Responding to the need and interest of the trans-national community they strive to represent, undoubtedly accounts for their tendency to opt for coherence and seamlessness, in spite of the fact that motivations for departure were diverse, and, in some cases, are detrimental to the communal spirit sought to be created in modern days; the shifts and silences found in the MoI's treatment of the Clearances are a good illustration of this. More generally, museum narratives in their interpretations of the Clearances, increasingly feature cameos of the fate of emigrants as shown for instance in the displays of Arran HM or Ullapool M&VC, and, perhaps more tellingly, in recently refurbished exhibitions which incorporate much more extensive coverage of 'remote connections' – the expression borrowed from Timespan's new exhibition.

Figure 9.2 Timespan, Helmsdale, Sutherland
Source: Laurence Gouriévidis 2008.

Discussing the ability of museums to communicate post-national, trans-cultural or 'hybrid' identities in late modernity, Macdonald concluded that 'museums are capable of being put to work in the expression of other kinds of identities than the national, homogeneous and bounded' and that they 'have long made connections between continents and between times' (S. Macdonald 2003: 10-11). Although dealing with other concerns than those tackled in this study, her point nevertheless emphasises the inventiveness and elasticity of exhibitions and their makers' vision re-orientating narratives. In the Clearances case, this manifests itself in the attempt to articulate trans-national identities through narratives of migration following the fate of Highland and Island emigrants in new homelands. However, as illustrated with the MoS, this approach may well have contributed to essentialising narratives of clearances emigration around, on the one hand, victimhood through the tropes of expulsion and exile, and on the other celebration of the contribution of Scottish emigrants abroad. The following section will consider the reverse gaze found in museum exhibitions presenting Highland immigrants to the New World, namely Australia.

Migrating memories: Clearance narratives in Australian migration museums

Immigration is at the core of definitions of Australian-ness, of the making of Australia and its social fabric, implicating notions of origins, foundational watersheds, colonisation and dispossession, inclusion and exclusion linked with immigration policies, acculturation and cultural imperialism. The weaving of such a host of cultural processes into a national narrative has followed a long ideological journey ranging from tales of discovery, settlement and birth of a new largely 'pom' state that whitewashed Indigenous populations and their territorial claims out of existence, to tales which reinscribed native voices alongside the plurality of immigrant accents and experiences.

The emergence of migration museums in Australia, roughly from the 1980s, is closely meshed in with the promotion of multi-culturalism as a dominant national discourse and policy, a promotion in which museums, as cultural agencies, were cast in key roles (McShane 2001: 122-25, Message 2009, Witcomb 2009). As a doctrine, cultural pluralism followed a long era of assimilationist policies and stance and seemed to herald new possibilities and visions – or revisions – of the past, but also new methodological and conceptual challenges for museum actors. In 1975 the Committee of Inquiry on Museums and National Collections published its report charting new directions for Australian museums. It invited a reassessment of the presentation of the past found in major Australian museums founded in the nineteenth century which 'tended to divorce Aboriginal man from European man and to divorce European man from nature' (Pigott 1975: 70). The Report recommended the division of museums into three grades: major, associated and local museums, conditioning federal expenditure, and also the establishment of a

Museum of Australia in Canberra which 'should not attempt to imitate or duplicate' the style of older Australian museums (ibid. 3-4). At the same time, as seen in Chapters one and six, elsewhere, the museological world was also experimenting with new approaches, in which social history and community participation were but two of the blades propelling change. In Australia, multi-culturalism as a novel concept championed differences and tolerance of others and implied a reversal of past governmental policies. This enabled museums to engage critically with long-ingrained readings of the past, in particular linked with the colonisation of Australia and the effects of immigration on Indigenous peoples and with the nature of Australian society, as well as introduce questioning and alternative perceptions and try new methodologies. Both migration museums studied are examples of these museological and ideological trends and developments in late twentieth-century Australia: the Migration Museum (MM) in Adelaide, South Australia, and the Immigration Museum (IM) in Melbourne, Victoria.

Adelaide's MM was the first of the two to be open to the public in 1986 and paved the way for this type of exhibitions in Australia. It is one of the museums funded by the State Government, which compose the divisions of the History Trust of South Australia. The Trust itself, a statutory authority reporting to Parliament through the Minister for the Arts, was established by the Government of South Australia in 1981 following a report by Robert Edwards, *Museum Policy and Development in South Australia* and a survey of the needs of South Australia's ethnic groups. The relativity of historical knowledge, the significance of changing and contradictory viewpoints underpin the Trust's approach to history laid out in guiding principles meant to assist staff; for instance they state:

- History can be interpreted and presented in many ways depending on the information available and the values and views of those presenting it. Therefore historical knowledge is often provisional and statements about the past may be hypotheses rather than infallible pronouncements …
- History bears witness to the past and must therefore address contentious and contested issues even when they make us uncomfortable.[3]

Such a vision of history runs through the MM's conception, structure and use of place.

The past functions of its very site are interpreted so as to shed light on the circumstances of the settlement of South Australia and immigration to the province and their implications. From 1845 to 1852, it was the location of an Aboriginal school where native children, forcibly removed from their families, were trained to learn the ways and values of white colonists. Then it became a destitute asylum where the aged, the sick, the orphans, but also unmarried mothers along with deserted wives and children – many of them recent immigrants – came

3 http://www.history.sa.gov.au/history/about_us/principles_history.htm [accessed 22 October 2008].

as a last resort to seek shelter behind its high walls, showing that life in the New World could also be, for some, synonymous with deprivation and exclusion. If the damaging effects of European settlement on Aboriginal peoples inseparable from the history of this building are told as part of the permanent exhibition in 'Behind the Wall', they are also acknowledged near the museum entrance by a 'Reconciliation Plaque' unveiled in 1999.

Figure 9.3 Reconciliation Plaque, Migration Museum, Adelaide, South Australia
Source: Laurence Gouriévidis 2005.

Both the plaque and a contemporary art installation by the Aboriginal artist Darryl Pfitzner Milika entitled 'Impact: An Illustrated History of Colonisation' greet visitors to the museum and introduce them to its perspective. The exhibition is chronologically structured gradually moving from exodus and early settlement to the twentieth century. During their journey through time visitors are repeatedly alerted to historiographical issues; questions are asked, phrasing is teased out and the contingency of values is prodded. For the foundation of South Australia, a series of panels entitled: 'Colonisation or invasion?' are placed in a narrow corridor graphically symbolising the squeezing out of Indigenous culture. Whilst the museum strives to discard the representation of Australia as *terra nullius*, it also strives to reflect the diversity of migrants' experiences and origins. The inclusion of a Forum Gallery, a dedicated space made over to community groups for three months at a time, is an attempt at translating into practice the ideal of mirroring a plurality of voices and identities. There, various groups can mount their own display and express their histories and cultures in ways specific to them and with limited input on the part of museum staff.

Incorporating a programme of Community events and exhibitions is a trademark of both Adelaide's and Melbourne's museums. In Melbourne, the IM hosts both community exhibitions and festivals and its website emphasises community engagement. In fact, this input from communities is seen, by some scholars, as 'probably one of the strengths of Australian museums' (McShane 2001: 131). The IM was initiated by the Victorian State Government and was opened in 1998, one of four venues in the care of Museum Victoria, a public museum organisation established in its present form in 1983, under the Museums Act of the same year. According to IM's manager, its creation followed a period of intense lobbying from various communities and almost a decade of feasibility studies (Sebastian 2007: 152).

Like the MM, it makes full use of its location, Old Customs House, one of Melbourne's most significant public buildings, symbolising trade and shipping and the economic expansion of both town and province. The building's function is integrated within a narrative of immigration structured around the migrants' experience with galleries entitled: Leavings, Settlings, Customs Gallery, Journeys, Getting in, each thematic section chronologically organised. The 'settlings' room also provides an overall background for the whole exhibition with a timeline detailing the peopling of Victoria from about the first European arrivals (not much is said of the continent's population before then) to the twenty-first century, emphasising legislative and event-based turning points, and, in demographic terms, contrasting state and national population. The multiplicity of experiences is reflected in the plurals used and the exhibition's subtitle 'moving stories' denoting personalised, but also poignant, biographies. The sharp contrast with the final gallery's singular form 'getting in' arguably signifies the rigidity of immigration policies filtering in selected groups.

Seemingly the theme resulted from 'market research conducted at the Museum that found that visitors wanted to know more about the process of getting in to Australia, what people went through, who was accepted and who wasn't, and the effect this had on society'.[4]

4 http://museumvictoria.com.au/getting-in [accessed 25 June 2009].

Figure 9.4 Immigration Museum, Melbourne, Victoria
Source: Laurence Gouriévidis 2005.

These two museums project the same ethos and belief. Both write Aboriginal Australia back into a narrative of population movement, confrontation and exchange; in both cases cultural diversity directs the museums' collecting, conservation and interpretive policies, as well as their education programs. But, above all, both are envisaged in terms of their social function as media disseminating values and eliciting questions with the potential of influencing attitudes. In the words of Padmini Sebastian (2007: 152), manager of the IM:

> The role of museums such as the Immigration Museum is important in connecting with communities and reflecting the diverse society that we live in. It is also a forum to "sell" the benefits of migration and diversity. Through dialogue and discussion, there are opportunities to inform and influence the character of the place that we live in, to encourage ongoing community discourse, and play a transformative role in society.

Likewise, as its core values, Adelaide's MM underlines 'social inclusion, countering racism, and life long learning'.[5] In terms of method, although structured differently, both museums juggle between on the one hand contextualising migratory waves and defining policies over time, and on the other, presenting individuated and localised stories personalising and humanising the representation. In this they reflect the tensions faced by many of the Highland and Island museums surveyed and more widely by museums with a social approach of the past: how to convey the local and personal whilst also imparting a sense of broader historical trends.[6]

The migration of Highlanders during the Clearances is one of many 'stories' illustrative of wider trends. Highland emigrants were only a fraction of the emigrants who arrived in Australia in the nineteenth century and only a fraction of those who came from the United Kingdom. Eric Richards has estimated that, in the nineteenth century, of European immigrants to Australia, the Scots formed about 13 per cent compared to the English 53 per cent and the Irish 23 per cent; the rate for Victoria was 15.7 per cent and for South Australia 11 per cent (Victoria and Queensland being the two states were a greater concentration of Scots were found) (Richards 1988: 12-13). Additionally, at its peak from the 1830s to the 1850s the emigration of Highlanders and Islanders to Australia seems much more uniform in terms of motivation, social origin and context. Most of those who left were amongst the poorest; they went at times of extreme hardship in the Highland and Island region, travelling with the help of government-assisted schemes such as, between 1837 and 1842, the bounty immigration scheme subsidised by the colonial government and ended during the colonial economic recession; between 1852 and 1857 in response to the great famine crisis, the Highland and Island Emigration Society (HIES) was engineered by Sir John McNeill, Chairman of the Board of Supervision of the Scottish Poor Law and Charles Trevelyan, Assistant Secretary to the Treasury, and combined colonial, philanthropic and proprietorial finances, the cost of the passage to be repaid by the recipient after arrival (Richards 1982-83, 2001, Devine 1988: 245-72, Harper 2003: 50-51). Over the period of operation of the HIES, about 4,910 people left for Australia, 3,134 going to Victoria and 848 to South Australia (Devine 1988: 260, Howe 1988a: 76-77). Museums tend to focus on those climactic moments – easily quantified through official statistics and illustrated with traceable examples of ships and passengers.

---

5 http://www.history.sa.gov.au/migration/about/history.htm [accessed 21 October 2008]; see also Szekeres (2005). Viv Szekeres was Director of MM between 1987 and 2008.

6 See Hutchison's article (2009) in which she discusses the alternative approaches to exhibiting migration history explored in a project run jointly by the National Museum of Australia and the Australian National University between 2005 and 2008. Her discussion set within the framework of discourses surrounding Australian diversity outlines previous exhibitionary models and presents the theory and practice underpinning the *Migrating Memories* exhibitions born out of the research undertaken.

Figure 9.5 Immigration Museum, Melbourne, Victoria
Source: Laurence Gouriévidis 2005.

Not surprisingly then, given its relative homogeneity, Clearance emigration to Australia lends itself to iconic representations standing for the flight of the poor and desperate in search of a better life. At the MM in Adelaide, Clearance emigrants are used alongside Famine emigrants from Ireland as the epitome of those who were escaping destitution and came to Australia 'in the spirit of hope'. Although this part of the exhibition introduces the broader ideology which underpinned the peopling of South Australia, in particular Wakefield's theory of systematic colonisation hinging on the combination of land, capital and labour – land revenues being used to fund the cost of emigrant passage,[7] governmental policies targeting British paupers are not mentioned, notably the schemes of assistance in place for the Irish and the Highlanders. The Scots, as one of the ethnic groups constituting the 'pioneers', feature later on in the exhibition and Gaelic-speaking Scots are clearly identified nuancing the sense of an all-encompassing Scottishness. Significantly, the construction of the 'pioneers' – as a seamless group – in collective memory is central to the exhibition, showing the socially conditioned nature of group

7 Wakefield also argued that the price of colonial lands had to be set sufficiently high to ensure that newcomers remained labourers and prevent the dispersion of settlers (S. Macintyre 2004: 78).

perception: 'Their struggle to succeed in a harsh environment made them heroes to later generations', followed in a later part by: 'Many aboriginal people lost their lives in violent conflict with the pioneers over the newcomers' occupation and use of land. Would you say that those who died were heroes or victims?' At the MM, as at the MoS, where the Clearances appear in abridged and condensed formats, they are used as a signifier for painful uprooting, emblematic of a set of pressures varying according to the perspective adopted; at the MoS,[8] they stand for coercive expulsion, and at the MM for the flight of the destitute. In Australia however, any form of celebratory narrative of rebirth and migrant contribution is tempered with the notion of cultural 'impact' and its undertones of damaging collision.

The Highland case is very specifically read in those terms at the IM. As mentioned Victoria was a prime Scottish and, in particular, Highland destination for migrants and Scottish 'stories' are among those particularised early on in the 'settlings' section. Niel Black's example and Scottish ceremonial artefacts form the backdrop and the visual frame through which the story of Scottish migration is apprehended (Figure 9.5).

Black arrived in 1839 – from Argyll although this is not specified – and, in his quoted words, the region was 'a Scotch settlement, and the people are so far as I can judge altogether Scotch'. He is the paradigm of a successful agricultural entrepreneur building his estate 'Glenormiston' 'into one of the wealthiest properties in the colonies'. But his experience is also used in sharp counterpoint to that of other fellow Scots whom he brought over to employ on his farm making use of the bounty scheme. As far as migration narratives go, the Scottish example is laced with contrasts but also heavy qualification. Contrast is found between compulsion and freedom of choice as 'entwined in Niel Black's story' are 'people [who] were forced to leave their homes, due to famine, oppression, disease and threat to their lives', their case illustrated with Ross' engraving for *Real Scottish Grievance* – undoubtedly a favoured image in major museums (see Chapter 4). Black's own entrepreneurial success is also qualified because it is shown to be based on the availability and import of an impoverished and pliable labour force. However, the most compelling qualification lies in the colonists' appropriation of Aboriginal land encapsulated in the duality of the title 'homeland gained, homeland lost'. In this section similarities are highlighted between the plight of Highland Scots and Aboriginal people 'equally displaced'; significantly, to Aboriginal displacement are also added conflict and brutality: 'there are many stories of intimidation by colonials of the Aborigines, of threats, poisonings, injury and killings'. Although the wording is more graphic than that depicting the clearing process in Scotland, they both resonate with the same sense of persecution. The Scottish case at the IM represents in microcosm the wider Anglo-Celtic colonisation of Australia and its long-lasting shock waves; the messages conveyed are mixed. Scots Australian seem in equal measure the agents and victims of the socio-economic ethos conditioning

8 As shown in Chapter 4, at the MoS, the Clearances are inserted in the narrative of Scottish emigration.

a vision of land which revolved on private ownership and the production of capital returns; it was a vision antithetical to that of Indigenous people resulting in their being persecuted and, ironically, one also used in Scotland in historicist discourse arguing in favour of communality of landholding in the Highland and Island region. Additionally the representation of Scottish immigration is also emblematic of the issues faced by Australian museums tackling social history: conveying a sense of the plurality of voices and experiences, in particular with respect to Anglo-Celtic migration and settlement.

Interestingly this was not a prime concern of another exhibition hosted in the Forum-Community Access Gallery at the MM in Adelaide: 'Scottish Migration to South Australia'. This temporary community event was organised in 2003 by the Scottish Associations of South Australia (SAoSA), two years before my visit to Australia.[9] To the interpretive community at work for this exhibition multiculturalism was not the central tenet underpinning selection of themes and perspective, neither was the reflexive approach – another tangible feature of the MM's permanent exhibition. Their prime concern was to portray Scottish emigration and Scots in South Australia, however their textual presentation was limited to the nineteenth-century. What was equally important to organisers was to impart to visitors a sense of their Scottish heritage and roots. A large part of the exhibition's text and artefacts therefore reflected their vision of the old 'homeland' and its culture. The panels presented the following themes:

- The origin of the Scottish people
- Scottish clans and their tartans
- The fight for independence – New Worlds
- Journey to the South Australian Colony
- Prominent Scots in early South Australia
- Scottish influence in South Australia.

Both this thematic division and the majority of artefacts – Robert Burns' memorabilia, displays revolving around pipe bands, Scottish country dancing and specific tartans – showed the shaping of a Scottish identity revolving on clanship and festivals amongst expatriate Scots and their descendants, an identity which owed much to the propagandist role of Caledonian and clan societies impelled by the desire to foster pride in Scottish roots and counter dilution (Howe 1988b: 114-19, MacLennan 2001: 49-50).

9 Although I did not see the exhibition, I met the organisers who made available to me the panels used and a wealth of photographic and other material. I am grateful to them for their assistance and their time.

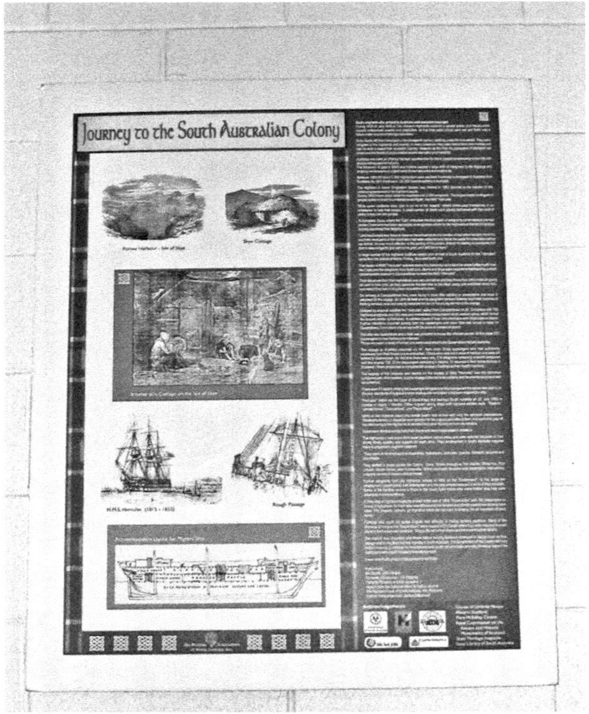

Figures 9.6 and 9.7 Migration Museum, Adelaide, Victoria: Display panels for the SAoSA's temporary community exhibition, 2003

Source: Laurence Gouriévidis 2005.

The reading of Scottish emigration by the SAoSA brought to light two parallel narrative models: the narrative of 'enrichment' and the narrative of painful uprooting. The immigrants' place in Australian society tended to be interpreted in conformity with long-standing Australian migration narratives such as that of 'enrichment' (McShane 2001: 128) or simply beneficial contribution, which was the focus of the last two panels listing the names and short biographies of over a score of famous or noteworthy Scots. The perspective adopted gave in small measure a flavour of what the *Scottish Reference Book*, published by the SAoSA a few years prior to the exhibition, had attempted. This was a memorial book meant, as its foreword outlined, to be 'a comprehensive record of our colourful culture and our expertise, which has helped to build this wonderful State of Australia … [and] to perpetuate the importance of our proud history' (Medhurst 1999: 6). Interestingly, in the exhibition, framing this collection of individual stories was a distinction, explicitly drawn in the final panel, between Highland Scots who 'came with few possessions and skills other than their family heritage and personal fortitude' and Lowland Scots who 'lived in a region with a culture of greater awareness of business and industry and fertile agricultural land. *They had a far better chance of achieving success in their new country*' (my emphasis). If Lowland Scots tended to connote success, the Highland example was well suited to the narrative of painful uprooting.

The Highland case was nevertheless pivotal in the SAoSA's narrative. It is through the Highland experience that the correlation between the two notions 'independence' and 'new worlds' was upheld in the section 'The fight for independence – New worlds', brought to life with examples of clearances predating the potato famine. Conflating, as this panel did, Bannockburn, Culloden and Sutherland removals was far from unambiguous; 'independence' seemed to be linked with political intentions rather than socio-economic motives and the context of nineteenth-century Scottish emigration, an integral part of British imperial policies and therefore Union, was left undefined. It was also through the Highland experience that the trope of the journey – another prominent feature of Australian migration exhibitions (McShane 2001: 129) – was tackled, most notably the journeys undertaken by cleared tenants (Figures 9.6 and 9.7). In fact the 'Journey' section of the display was almost entirely devoted to the work of the HIES, in particular relating at length the voyage of H.M.S. *Hercules*, which had received wide press coverage,[10] and mentioning subsequent vessels. In effect, the SAoSA's presentation of famine-driven emigrations to the New World during the Clearances is one of the most extensive of all exhibitions featured in this book. The story of the 'refugees' of the Highlands and Islands – borrowing from wider modern grammars – certainly underscored the tragedy of famine in the western Highlands, of the passage on the *Hercules* plagued by disease and death, and of

10 Against optimistic forecasts, the voyage of the *Hercules*, a ship lent by the Board of Admiralty, was dogged by adversity and turned into a protracted ordeal lasting over 100 days.

the fate of those of the HIES emigrants who reached Australia when employment opportunities had dried out and found themselves in a similar predicament as that which had motivated their departure from Scotland. The situation was summed up in stark terms: ‘it was very much a case of the survival of the fittest’. If filtering through the narrative were the requirement for emigrants to settle up with the HIES after arrival, or the interlocking conditions between Scotland and the colony affecting the emigrants’ ease or difficulty of absorption in their new homeland and necessitating community and family dispersal, the decidedly utilitarian purpose of the HIES was not explained.[11] In the story, the socio-economic context was balanced against imparting a sense of the emotional implications of the process of emigration, a process presented as distressing for the impoverished tenantry of course, but also for a ‘small number of chiefs’ such as Lord Macdonald of Skye. He was described, in the words of an HIES officer, as ‘much affected at the parting of his people; indeed his feelings so overcame him that he was obliged to give vent to his grief and I left him in tears’ – a scene and quote that no Scottish museum ever echoes and that lend credence to interpretations foregrounding the remnants of an underlying sense of kinship along with the ineluctability of the situation. The selection of this small vignette exemplifies the mechanism of remembrance with its ambiguities, shortcuts and ultimately bases in modern social dynamics and locale. Another important facet of migration conveyed in this community exhibition was that arrival in Australia did not mark the end but the continuation of a long journey forcefully symbolised with a trunk. ‘Tales from a trunk’ recounted Janet Fraser’s journey from Dornoch to Port Adelaide then on to various parts of South Australia and were penned by her great grand-daughter. Through this artefact and its tales was imagined the ‘pioneers’ experience: their isolation, their fear of Aboriginal population, and their struggle against a harsh environment where bushfires were a fact of life – in a nutshell, a conventional narrative of survival in hostile surroundings and adaptation.

This conventional narrative is somewhat in keeping with the tone found in the MoS’ representation of Scots abroad; common trans-national master-narratives of Clearance emigration therefore appear to be uniting painful uprooting and pioneering struggles. Judging from the weight given to Highland and Island emigrants in the SAoSA’s exhibition, it also seems that the Clearances play a disproportionate part in the collective memory of expatriate Scots and their descendants. It does appear that to them the Clearances ‘emerge as the historical trauma through which the existential anxieties of people of Scottish or part-Scottish descent dispersed throughout the world may be acted out, narrated and brought into the public domain’ (Basu 2007: 200) coupled with, more prosaically, the fact that sources may well be more easily traceable and available for displays

11 It was designed to solve, at one and the same time, the malfunction of the Poor Law in Scotland, the need for labour in South Australia and the problems of a burdensome tenantry for the landlords, whilst, ideally and philanthropically, giving the removed paupers the opportunity to help and reform themselves; see Richards (1993).

than those for the steady trickle of other migrants. Yet the appeal of clanship and surrounding folklorisation widely promoted by such associations as Caledonian Societies, initially meant to elicit some sense of Scottishness in societies increasingly becoming ethnically-mixed causes interferences in the perception of this historical trauma. This vision stemming from an old conservative order does not always quite tot up with the reformist vision promoted overall in Highland and Island public memory in Scotland – one in which the old patriarchs are often clad in the garb of promoters of expulsion. In museum exhibitions, present issues of power, commerce and identity politics – often at variance with each other – surface through their symbolic negotiations; such issues are located in time and place (Ashworth and Graham 2005: 3-12). Summoned up as a resource for each aspect, different facets of the past may simultaneously be brought to light.

Appendix:
Museum/Heritage Centre Questionnaire

MUSEUM/HERITAGE CENTRE QUESTIONNAIRE

NAME: **POSITION:**

1. Details

- When was the museum/centre first opened?

2. Origin of museum/centre

- What is the origin of the museum/centre i.e. trust, historical society, private individual?
- What was the initial motivation behind its creation?

  □ Absence of information locally
  □ Neglect of Highland/Island history
  □ Economic potential
  □ Increasing importance of heritage interpretation
  Other – please specify:

- Have changes/refurbishments taken place since its opening?
  If so, please specify:

3. Content of the display

- Timespan covered:

  □ Restricted period
  □ Development/changes in long term perspective

- Is there a running theme through the display? (Please expand on its nature)

- How were the topics and/or overall idea chosen?

- What is the basis of the presentation?

  □ Objects
  □ Audio-visual means
  □ Text
  □ Reconstructed scenes
  Other – please specify:

- Why was such a presentation chosen?

- What is its aim?

  □ Tell about people's life in the past
  □ Tell about the major events which shaped the region
  □ Tell about the culture of the region
  Other – please specify:

- Is there an attempt to link past events with the contemporary Highland/Island situation? (Please expand)

4. Creators of the display

- Who organized and thought out the content of the display?
- Were the organizers from the Highlands/Islands?
- Were they Gaelic speakers?
- Was anyone consulted?

  □ Historian (please give name(s))
  □ Local community
  Other – please specify:

- Were any official/professional guidelines referred to when the display was designed? (Please expand)

5. Recipients

- Who is your targeted audience?

  □ Visitors
  □ Local population

- Are visitors' surveys carried out on the impact of the exhibition? (Please expand)

6. Finance

- How is the museum/centre funded? Please list the sources of funding.

- Are there any conditions attached to funding i.e. topics to be included in the exhibition; format of the display; financial return expected; specific specialist recommended for consultation concerning the historical content?

7. Role of the museum/centre

- What are the main objectives of the museum/centre? (Answers may be classified (1. 2. 3. …))

  □ Contribution to the dissemination of Highland and Island culture
  □ Resource centre
  □ Teaching or informing visitors about Highland and Island history
  □ Tourist attraction
  □ Make a profit
  □ Contribution to the local economy i.e. jobs
  (Please give some details here)

  Other – please specify:

- Which of the following criteria were considered when the exhibition was planned? (Answers may be classified)

  □ Accuracy
  □ Wide appeal
  □ Authenticity
  □ Sense of local community and culture
  □ Entertainment
  Other – please specify:

- Is the exhibition part of a complex offering other services?

  □ Restaurant
  □ Art gallery
  □ Gift shop
  Other – please specify:

- If so, why?

8. Marketing

- Is the museum/centre a member of any official bodies? (Please give details)

9. Social impact

- Can the museum/centre be described as a museum/centre of the local community?

  □ Involvement with schools
  □ Temporary community exhibitions
  □ Involvement with local history society
  Other – please specify?

10. The Clearances and/or the Land War

- How important is the period of the Clearances and/or the Land War in the display?

- Why was it included?

- How is it presented?

  □ Pictures/paintings
  □ Text
  □ Artefacts
  □ Audio-visual means
  □ Reconstructed scene(s)
  Other – please specify:

- Please explain this choice of presentation:

- Which sources were consulted? (Please give details if possible)

  □ Contemporary sources
  □ Oral history
  □ Modern interpretations
  Other – please specify?

11. Interpretation of Highland history

- Was any particular perspective adopted in the presentation of this period?

  □ The improvers' viewpoint?
  □ The crofting community's viewpoint?
  Other – please specify:

- If so, why?

- How much attention was paid to modern history writing on the period and why?

- There is not always widespread agreement on the interpretation of the period. Is it portrayed as a contested issue in the display? (Please explain)

- How important is the question of 'objectivity' in such a presentation? Can, in fact, museums/centres adopt an ideological position?

- What would you say is the importance of the period in Highland/Island history and culture?

- Is it shown, in the exhibition, as having shaped the present day Highlands/ Islands? (If so, please explain)

- We hear quite a lot about the question of Scottish identity, how important do you think the period is to Highland/Island identity?

I would welcome any other comments that you may wish to make.

Bibliography

Adam, R.J. 1972. *Papers on the Sutherland Estate Management 1802-1816*. 2 vols. Edinburgh: Scottish History Society.

Advisory Council on Education in Scotland. 1951. *Libraries, Museums and Art Galleries* (Cmd. 8229), HMSO.

Aitchison, P. and Cassell A. 2003. *The Lowland Clearances: Scotland's Silent Revolution 1760-1830*. East Linton: Tuckwell.

Aldridge, D. 1973. Museums and historic interpretation with reference to an ideal national pattern and the Scandinavian and American analogies, in *Country Life Museums: Report on a Conference Held in January 1972*, edited by the Countryside Commission for Scotland and Scottish Country Life Museums Trust. Perth: Countryside Commission for Scotland, 32-38.

Alexander, E.P. 1979. *Museums in Motion: An Introduction to the History and Functions of Museums*. Nashville: American Association for State and Local History.

Ambrose, T. 1989a. Preface, in *Presenting Scotland's Story*, edited by T. Ambrose. Edinburgh: Scottish Museums Council, HMSO, 5.

Ambrose, T. 1989b. Local authorities and museums. *Scottish Museums News*, Winter, 1.

Ambrose, T. 1990a. Private sector support for museums. *Scottish Museums News*, Summer, 1.

Ambrose, T. 1990b. Tourism and museums. *Scottish Museums News*, Winter, 1.

Ambrose, T. 1992a. European connections. *Scottish Museums News*, 8(3) Autumn, 1.

Ambrose, T. 1992b. Museums and the natural environment. *Scottish Museums News*, 8(4) Winter, 1.

Ambrose, T. 1996. The Scottish Museums Council: A model of museum support. *Museum International*, 48(3), 43-46.

Anderson, B. 1991. *Imagined Communities*. First published 1983. London: Verso.

Anderson, J. 1785. *An Account of the Present State of the Hebrides and Western Coasts of Scotland*. Edinburgh.

Anderson, R. 1989. Scotland's history in the National Museums of Scotland, in *Presenting Scotland's Story*, edited by T. Ambrose. Edinburgh: Scottish Museums Council, HMSO, 65-74.

Anon. 1995. In praise of Henry George, in *Tuath is Tighearna. Tenants and Landlords*, edited by D.E. Meek. Edinburgh: Scottish Academic Press, 240.

Ansdell, D. 1998. *The People of the Great Faith: The Highland Church 1690-1900*. Stornoway: Acair.

Antze, P. and Lambek, M. 1996. *Tense Past: Cultural Essays in Trauma and Memory*. London: Routledge.
Ardrossan and Saltcoats Herald. 1976. Open day at museum, 20 August, 6.
Ardrossan and Saltcoats Herald. 1977. Museum plans unveiled, 11 March, 6.
Ardrossan and Saltcoats Herald. 1979. Opening of the museum fulfils 10-year dream, 20 July, 4.
Aros. n.d. *Aros. Skye's Heritage Centre. Dualchas an Eilein*. Portree: Pelican Design Consultants.
Ascherson, N. 2002. *Stone Voices. The Search for Scotland*. London: Granta.
Ashplant, T.G., Dawson, G. and Roper, M. 2004a. *Commemorating War: The Politics of Memory*. New Brunswick, NJ: Transaction Publishers.
Ashplant, T.G., Dawson, G. and Roper, M. 2004b. The politics of war memory and commemoration: Contexts, structures and dynamics, in *Commemorating War: The Politics of Memory*, edited by T.G. Ashplant et al. New Brunswick, NJ: Transaction Publishers, 3-85.
Ashworth, G.J. and Graham, B. 2005. *Sense of Place: Sense of Time*. Aldershot: Ashgate.
Assman, A. 2004. Four formats of memory: From individual to collective constructions of the past, in *Cultural Memory and Historical Consciousness in the German-Speaking World since 1500*, edited by C. Emden and D. Midgley. Oxford: P. Lang, 19-37.
Association for Scottish Visitor Attractions. 1991. *News*. March.
Atkinson, F. 1972. Open air museums and environmental conservation. *Museums Journal*, 72(3), 97-99.
Atkinson, F. 1975. Presidential address. *Museums Journal*, 75(3), 103-105.
Atkinson, F. and Holton, M. 1973. Open air and folk museums. *Museums Journal*, 72(4), 140-42.
Attwood, B. 2004. Whose dreaming? Reviewing the *Review of the National Museum of Australia*. *History Australia*, I(2), 279-92.
Aughey, A. 2001. *Nationalism, Devolution and the Challenge to the United Kingdom State*. London: Pluto Press.
Austin, R. 1958. Scandinavian Folk Museums. *Museums Journal*, 57(11), 253-57.
Australia ICOMOS. 1999. *The Burra Charter: The Australia ICOMOS Charter for the Conservation of Places of Cultural Significance*.
Baillie, D. 1995. Satire on Patrick Sellar, in *Tuath is Tighearna. Tenants and Landlords*, edited by D.E. Meek. Edinburgh: Scottish Academic Press, 190-1.
Baker, E. 1923. *The Highlands with Rope and Rucksack*. London: H.F. and G. Witherby.
Balfour, H. 1904. The relationship of museums to the study of anthropology. *Museums Journal*, 3(12), 396-408.
Balfour, H. 1909. Presidential address. *Museums Journal*, 9(1), 5-18.
Balfour, H. et al. 1912. A national folk museum. *Museums Journal*, 11(8), 221-25.
Basu, P. 2000. Sites of memory – sources of identity: Landscape-narratives of the Sutherland Clearances, in *Townships to Farmsteads: Rural Settlement Studies*

in Scotland, England and Wales, edited by J.A. Atkinson, I. Banks and G. MacGregor. Oxford: John and Erica Hedges, 225-236.
Basu, P. 2007. *Highland Homecomings: Genealogy and Heritage Tourism in the Scottish Diaspora*. London: Routledge.
Bather, F.A. 1903. Presidential address. *Museums Journal*, 3(3), 71-94.
Bather, F.A. 1911. Open-air folk-museums: Progress and prospects. *Museums Journal*, 10(9), 249-53.
Bather, F.A. 1916. The triumph of Hazelius. *Museums Journal*, 16(6), 132-36.
Bell, J. 1996. Making rural histories, in *Making Histories in Museums*, edited by G. Kavanagh. London: Leicester University Press, 30-41.
Bennett, T. 1988. Museums and 'the people', in *The Museum Time-Machine. Putting Culture on Display*, edited by R. Lumley. London: Routledge, 63-85.
Bennett, T. 1995. *The Birth of the Museum: History, Theory, Politics*. London: Routledge.
Bitterman, R. 1999. On remembering and forgetting: Highland memories within the maritime diaspora, in *Myth, Migration and the Making of Memory: Scotia and Nova Scotia c. 1700-1990*, edited by M. Harper and M.E. Vance. Edinburgh: John Donald and Halifax: Frenwood Publishing, 253-265.
Borley, L. 1988. Museums and tourism, in *Working with Museums*, edited by T. Ambrose. Edinburgh: HMSO, 11-17.
Boswell, D. and Evans, J. 1999. *Representing the Nation: A Reader. Histories, Heritage and Museums*. London: Routledge.
Boylan, P. 1990. Museums and cultural identity. *Museums Journal*, 90(10), 29-33.
Brand, J. 1978. *The National Movement in Scotland*. London: Routledge.
Broad Oak Consultancy. 1993. *A Report on Museums Development in the Highlands and Islands of Scotland. Prepared for HIE, SMC and HIMF*. Heathfield: Broad Oak Consultancy.
Bruce, J. 1847. *Destitution in the Highlands. Letters on the Present Condition of the Highlands and Islands of Scotland. Reprinted from the Scotsman Newspaper*. Edinburgh: Adam Charles and Black.
Bumsted, J.M. 1982. *The People's Clearance: Highland Emigration to British North America 1770-1815*. Edinburgh: Edinburgh University Press.
Calder, C. 1993. Tourist centre shows Skye is no limit – Scotland. *Sunday Times*, 5 December.
Calder, J. 2000. From artefacts to audience. Strategy for display and interpretation, in *Heritage and Museums. Shaping National Identity*, edited by J.M. Fladmark. Shaftesbury: Donhead Publishing, 41-52.
Callander, R. 1998. Still living under feudalism. *Scotland on Sunday*, 26 April, 15.
Cameron, E.A. 2001. 'Unfinished business': The land question and the Scottish parliament. *Contemporary British History*, 15(1), 83-114.
Cannizo, J. 1991. Exhibiting cultures: 'Into the heart of Africa'. *Visual Anthropology Review*, 7(1), 150-160.
Carmichael, M.C. 2002. *Dualchas* – Skye and Lochalsh Area Museums Service, the Highland Council. *Scottish Archives*, 8, 113-120.

Carrington, L. 1999. An open door policy. *Museums Journal*, 99(7), July, 26-29.

Carroll, J. 2004. A response to Bain Attwood's 'Whose drealing?' *History Australia*, I(2), 293-94.

Carter, I. 1971. Economic models and the recent history of the Highlands. *Scottish Studies*, 99-120.

Carter, I. 1981a. The changing image of the Scottish peasantry 1745-1980, in *People's History and Socialist Theory*, edited by Raphael Samuel. London: Routledge.

Carter, I. 1981b. The Scottish peasantry, in *People's History and Socialist Theory*, edited by Raphael Samuel. London: Routledge.

Casey, D. 2001. The National Museum of Australia: Exploring the past, illuminating the present and imagining the future, in *National Museums. Negotiating Histories*, edited by D. McIntyre and K. Wehner. Canberra: National Museum of Australia, 3-11.

Cheape, H. 1979. The Agricultural Museum, in *The Past and the Present. The Role of the National Museum of Antiquities of Scotland*, edited by A. Fenton. Edinburgh: National Museum of Antiquities of Scotland, 17-20.

Cheape, H. 1986. Dr I.F. Grant (1887-1983): The Highland Folk Museum and a bibliography of her written works. *Review of Scottish Culture 2*, 113-125.

Clark, H. 1988. Community involvement in 'The People's Story'. *Scottish Museums News*, Autumn, 4-6.

Clark, R. 1993. Museum registration in Scotland: Three years of progress. *Scottish Museums News*, 9(1) Spring, 18.

Clifford, J. 1999. Museums as contact zones, in *Representing the Nation: A Reader. Histories, Heritage and Museums*, edited by D. Boswell and J. Evans. London: Routledge, 435-457.

Clyde, R. 1995. *From Rebel to Hero: The Image of the Highlander 1745-1830*. East Linton: Tuckwell Press.

Cohen, A.P. 1989. *The Symbolic Construction of Community*. First published 1985. London: Routledge.

Conan, E. and Rousso, H. 1996. *Vichy, un passé qui ne passe pas*. Second edition. Saint-Amand: Gallimard Folio histoire.

Connerton, P. 1989. *How Societies Remember*. Cambridge: Cambridge University Press.

Coombes, A.E. 2004. Museums and the formation of national and cultural identities, in *Museum Studies: An Anthology of Contexts*, edited by B.M. Carbonell. Oxford: Blackwell, 231-246.

Corsane, G. 2005. *Heritage, Museums and Galleries. An Introductory Reader*. London: Routledge.

Coutts, H. and Clark, H. 1989. Telling the story. *Museums Journal*, 89(8), 30-33.

Cowan, E.J. 1977-8. Agricultural improvement and the formation of early agricultural societies in Dumfries and Galloway. *Transactions of the Dumfrieshire and Galloway Natural History and Antiquarian Society*, 3rd series, LIII, 157-67.

Cowan, E.J. 1992. From the Southern uplands to Southern Ontario: Nineteenth-century emigration from the Scottish borders, in *Scottish Emigration and Scottish Society*, edited by T.M. Devine, Edinburgh: John Donald, 61-83.

Craig, D. 1990. *On the Crofters' Trail: In Search of the Clearance Highlanders*. London: Pimlico.

Craig, D. 1991. *King Cameron*. Manchester: Carcanet.

Crane, S.A. 2006. The conundrum of ephemerality: Time, memory and museums, in *A Companion to Museum Studies*, edited by S. Macdonald. Oxford: Blackwell, 98-109.

Cregeen, E. 1974a. Oral tradition and agrarian history in the West Highlands. *Oral History*, 2(1), 15-33.

Cregeen, E. 1974b. Oral sources for the social history of the Scottish Highlands and Islands. *Oral History*, 2(2), 23-36.

Cregeen, E. 1993-1998. Oral tradition and history in a Hebridean Island. *Scottish Studies*, 32, 12-37.

Crofters' Commission. 1991. *Crofting in the 1990s*. Inverness.

Crooke, E. 2007. Museums, communities and the politics of heritage in Northern Ireland, in *Museums and their Communities*, edited by S. Watson. London: Routledge, 300-12.

Cubbon, A.M. 1972. A museum and its environment. *Museums Journal*, 72(3), 95-6.

Cultural Trends. 1989 (4). London: Policy Studies Institute.

Cultural Trends. 1991 (9). London: Policy Studies Institute.

Cultural Trends. 1995 (28). London: Policy Studies Institute.

Cultural Trends in Scotland. 1992. London: Policy Studies Institute in association with Scottish Arts Council.

Davidson, G. 2001. National museums in a global age: Observations abroad and reflections at home, in *National Museums: Negotiating Histories*, edited by D. McIntyre and K. Wehner. Canberra: National Museum of Australia, 12-28.

Davies, M. 1989. Heritage for profit. *Museums Journal*, 89(1), 17-8.

Davis, P. 1999. *Ecomuseums. A Sense of Place*. London: Leicester University Press.

Davison, P. 2005. Museums and the re-shaping of memory, in *Heritage, Museums and Galleries: An Introductory Reader*. London: Routledge.

Dawson, G. 2005. Trauma, place and the politics of memory: Bloody Sunday, Derry, 1972-2004. *History Workshop Journal*, 59, 151-78.

Dawson, J. 1998. The Gaidhealthachd and the emergence of the Scottish Highlands, in *British Consciousness and Identity: The Making of Britain 1533-1707*, edited by B. Bradshaw and P. Roberts. Cambridge: Cambridge University Press.

Devine, T.M. 1988. *The Great Highland Famine: Hunger, Emigration and the Scottish Highlands in the Nineteenth Century*. Edinburgh: John Donald.

Devine, T.M. 1989. Social responses to agrarian 'Improvement': The Highland and Lowland Clearances in Scotland, in *Scottish Society 1500-1800*, edited by

R.A. Houston and I.D. Whyte. Cambridge: Cambridge University Press,148-168.

Devine, T.M. 1992. Landlordism and Highland emigration, in *Scottish Emigration and Scottish Society*, edited by T.M. Devine. Edinburgh: John Donald, 84-103.

Devine, T.M. 1994a. *Clanship to Crofters' War: The Social Transformation of the Scottish Highlands*. Manchester: Manchester University Press.

Devine, T.M. 1994b. *The Transformation of Rural Scotland: Social Change and the Agrarian Economy 1660-1815*. Edinburgh: Edinburgh University Press.

Devine, T.M. 1999a. A conservative people? Scottish Gaeldom in the age of Improvement, in *Eighteenth Century Scotland: New Perspectives*, edited by T.M. Devine and J.R. Young, East Linton: Tuckwell Press, 225-37.

Devine, T.M. 1999b. *The Scottish Nation 1700-2000*. London: Penguin.

Devine, T.M. 2006. *Clearance and Improvement: Land, Power and People in Scotland 1700-1900*. Edinburgh: John Donald.

Dewar, D. 1998. *Land Reform for the Twenty-First Century: 5th John McEwen Memorial Lecture*, 4 September 1998. Available at: http://www.caledonia.org.uk [accessed 1 August 2007].

Dewey, C. 1974. Celtic agrarian legislation and the Celtic revival: Historicist implications of Gladstone's Irish and Scottish Land Acts 1870-1886. *Past and Present*, 64, 30-70.

DiCenzo, M. 1996. *The Politics of Alternative Theatre in Britain 1968-1990. The Case of 7:84 (Scotland)*. Cambridge: Cambridge University Press.

Dodgshon, R.A. 1998. *From Chiefs to Landlords: Social and Economic Change in the Western Highlands and Islands c. 1493-1820*. Edinburgh: Edinburgh University Press.

Donnachie, I. and Whatley, C. 1992. *The Manufacture of Scottish History*. Edinburgh: Polygon.

Dumoulin, O. 2003. *Le rôle social de l'historien. De la chaire au prétoire*. Paris: Albin Michel.

Edwards, R. 1981. *Museum Policy and Development in South Australia: Final Report*. Adelaide: Government Printer.

Empire Exhibition Scotland 1938. Bellahouston Park, Glasgow. Official Catalogue, 1938. Glasgow: Empire Exhibition.

Erskine of Mar, R. 1918a. The Russian constitution. *The Scottish Review*, 41(90), 265-6.

Erskine of Mar, R. 1918b. The Ulster Problem. *The Scottish Review*, 41(90), 269-71.

Fairhurst, H. 1967-68. Rosal: A deserted township in Strath Naver, Sutherland. *Proceedings of the Society of Antiquaries of Scotland*, 100, 135-169.

Farge, A. 1997, *Des lieux pour l'histoire*. Paris: Editions du Seuil.

Fenton, A. 1962. Scottish agricultural history museums. *Transactions of the Royal Highland and Agricultural Society of Scotland*, 6th series, vii, 54-62.

Fenton, A. 1973a. The Scottish Country Life Museums Trust, in *Country Life Museums: Report on a Conference Held in January 1972*, edited by the Countryside Commission for Scotland and Scottish Country Life Museums Trust. Perth: Countryside Commission for Scotland, 2-6.

Fenton, A. 1973b. Regional ethnology and environmental awareness. *Museums Journal*, 73(3), 107-10.

Fenton, A. 1978. The Scottish Agricultural Museum. *Museums Journal*, 77(4), 167-8.

Fenton, A. 1985. *The Shape of the Past 1: Essays in Scottish Ethnology*. Edinburgh: John Donald.

Fenton, A. and Sprott, G. 2002-2003. The Museums of Scottish Country Life. *Review of Scottish Culture*, 15, 3-12.

Fentress, J. and Wickham, C. 1992. *Social Memory*. Oxford: Blackwell.

Fenyo, K. 2000. *Contempt, Sympathy and Romance: Lowland Perceptions of the Highlands and the Clearances during the Famine Years 1845-1855*. East Linton: Tuckwell Press.

Fidler, K. 1984. *The Desperate Journey*. First published 1964. Edinburgh: Canongate.

Finlay, I. 1964. What image? What public? *Museums Journal*, 64(3), 248-53.

Finlay, R.J. 1994. *Independent and Free: Scottish Politics and the Origins of the Scottish National Party 1918-1945*. Edinburgh: John Donald.

Fladmark, J.M. 1994. *The Wealth of a Nation. Heritage as a Cultural and Competitive Asset*. Aberdeen: The Robert Gordon University.

Flinn, M.W. 1977. Malthus, emigration and potatoes in the Scottish north-west 1770-1870, in *Comparative Aspects of Scottish and Irish Economic and Social History 1600-1900*, edited by L.M. Cullen and T.C. Smout. Edinburgh: John Donald, 47-64.

Forestry Commission. 1995. *Forest and Archaeology: Guidelines*.

Forsyth, D. 2000. Scotland and the world. Building on the diaspora, in *Heritage and Museums. Shaping National Identity*, edited by J.M. Fladmark. Shaftesbury: Donhead Publishing, 115-122.

Fox, C. 1934. Open-air museums: Presidential address to the Museums Association. *Museums Journal*, 34(4), 109-21.

Franck, R. 2001. La France des années noires: La mémoire empoisonnée. *Cahiers Français*, 303, 56-65.

Fraser, G.M. 2005. *Patrick Sellar as a Young Man (1801-1810)*. Kinloss: Librario.

Fraser, M. 1982-1983. Countryside grant and museum developments. *Scottish Museums News*, Dec 1982-Jan 1983, 3-5.

Fry, M. 2005. *Wild Scots. Four Hundred Years of Highland History*. London: John Murray.

Fyfe, G. 2006. Sociology and social aspects of museums, in *A Companion to Museum Studies*, edited by S. Macdonald. Oxford: Blackwell Publishing, 33-49.

Gibson, R. 1978. It's Scotland soil. *Crann-tara* [online], 2, Spring 1978. Available at: http://www.Caledonian.org.uk [accessed 1 August 2007].

Gibson, R. 1996. *Toppling the Duke – Outrage on Ben Bhraggie?* Evanston: Highland Heritage Books.

Gibson, R. 2006. *The Highland Clearances Trail*. Edinburgh: Luath Press.

Gibson, R. n.d. *Highland Clearances Trail – A Guide*. Third edition.

Gouriévidis, L. 1993. *The Image of the Highland Clearances c. 1880-1990*. Unpublished PhD thesis. University of St Andrews.

Gouriévidis, L. 2000a. The Strathnaver Clearances in modern Scottish fiction: history, literary perception and memory, in *The Province of Strathnaver*, edited by J. Baldwin. Edinburgh: Scottish Society for Northern Studies, 195- 214.

Gouriévidis, L. 2000b. Agrarian changes and the Highland peasantry 1790s-1880s: Beneficient progress or victims' ordeal? *Ranam*, 33, 181-196.

Gouriévidis, L. 2001. Discours recial et politique socio-économique dans les *Highlands* au XIXème siècle. *Cahiers victoriens et édouardiens*, 54, 125-136.

Gouriévidis, L. 2003-2004. La racialisation des *Highlanders* et l'église Presbytérienne en Ecosse au milieu du XIXème siècle. *Etudes écossaises*, 9, 369-386.

Gouriévidis, L. 2004. La randonnée dans les *Highlands* durant l'entre-deux-guerres: Enjeux socio-économiques, in *Le loisir en Ecosse*, edited by J. Berton. St Etienne: Publications de l'Université de St Etienne, 135-145.

Gouriévidis, L. 2005. Patrick Sellar. *Etudes écossaises*, 10, 73-88.

Gouriévidis, L. 2008. Le présent du passé: La commémoration de l'histoire des *crofters* sur l'île de Lewis. *Etudes écossaises,* 11, 175-192.

Graham, J.J. 1987. *Shadowed Valley*. Lerwick: The Shetland Publishing Company.

Grant, I.F. 1929. The forthcoming Highland Exhibition. *The Scots Independent*, iv, 119-120.

Grant, I.F. 1930a. Foreword. *The Highland Exhibition 1930 Illustrating the History, Antiquities, Folk Life, Arts and Crafts, and the Romance of the Highlands and Islands from the Earliest Times to the Present Day – Catalogue*. Inverness, xxvi-xxxi.

Grant, I.F. 1930b. A sacrament of folk-life. *The Scots Magazine*, xiii (4), 283-86.

Grant, I.F. 1936. *Am Fasgadh*. A museum on the Isle of Iona. *Museums Journal*, 36(8), 347-8.

Grant, I.F. 1938. Highland homestead, in *Empire Exhibition. Scotland 1938. Scottish Pavilions. Official Guide*. Glasgow: Empire Exhbition, 50-1.

Grant, I.F. 1945. *Am Fasgadh. The Highland Folk Museum at Kingussie, Inverness-shire*. Glasgow: MacLehose.

Grant, I.F. 1962. Photographs of remembrance, *The Scots Magazine*, 77(5), August, 439-448.

Grant, I.F. 1989. *Highland Folk Ways*. First published 1961. London: Routledge.

Grant, I.F. 2007. *The Making of Am Fasgadh*. Edinburgh: National Museums of Scotland.

Grassie, J. 1983. *Highland Experiment: The Story of the Highlands and Islands Development Board*. Aberdeen: Aberdeen University Press.

Gray, M. 1957. *The Highland Economy 1750-1850*. Edinburgh: Oliver and Boyd.

Greenhalgh, P. 1988. *Ephemeral Vistas. The Expositions Universelles, Great Exhbitions and World Fairs 1851-1939*. Manchester: Manchester University Press.

Gregory, K. and Witcomb, A. 2007. Beyond nostalgia. The role of affect in generating historical understanding at heritage sites, in *Museum Revolutions. How Museums Change and Are Changed*, edited by S.J. Knell, S. MacLeod and S.E.R Watson. London: Routledge, 263-275.

Grimble, I. 1962. *The Trial of Patrick Sellar: The Tragedy of Highland Evictions*. London: Routledge.

Grimble, I. 1976. Patrick Sellar, in *History is my Witness*, edited by G. Menzies. London: British Broadcasting Corporation, 38-66.

Gunn, N.M. 1954. The Highland Clearances. *Radio Times*, 24 November. NLS, Dep 209, Box 8.

Gunn, N.M. 1966. The lingering problem of the Highlands. *The Glasgow Herald*, 10 December, 6.

Gunn, N.M. 1977. *Butcher's Broom*. First published 1934. London: Souvenir Press.

Gunn, N.M. 1987a. Belief in Ourselves, in *Landscape and Light. Essays by Neil M. Gunn*, edited by A. McCleery. First published in 1945. Aberdeen: Aberdeen University Press, 158-161.

Gunn, N.M., 1987b. Scotland moves, in *Landscape and Light. Essays by Neil M. Gunn*, edited by A. McCleery. First published 1943. Aberdeen: Aberdeen University Press, 162-165.

Halbwachs, M. 1980. *The Collective Memory*. New York: Harper and Row.

Halbwachs, M. 1992. *On Collective Memory*. Chicago: University of Chicago Press.

Halbwachs, M. 1994. *Les cadres sociaux de la mémoire*. 2nd edition. Paris: Albin Michel.

Halbwachs, M. 1997. *La mémoire collective*. 2nd edition. Paris: Albin Michel.

Halbwachs, M. 2008. *La topographie légendaire des évangiles en Terre sainte*. First published 1941. Paris: PUF.

Hanham, H.J. 1969a. *Scottish Nationalism*. London: Faber.

Hanham, H.J. 1969b. The problem of Highland discontent, 1880-1885. *Transactions of the Royal Historical Society*, fifth series, 19, 21-65.

Harden, D.B. 1956. Museum development in Great Britain. *Museums Journal*, 56(7), 161-165.

Harper, M. 2003. *Adventurers and Exiles: The Great Scottish Exodus*. London: Profile.

Harper, M. and Vance, M.E. 1999. *Myth, Migration and the Making of Memory: Scotia and Nova Scotia c. 1700-1900*. Edinburgh: John Donald and Halifax: Fernwood Publishing.

Harrison, J.D. 2005. Ideas of museums in the 1990s, in *Heritage, Museums and Galleries: An Introductory Reader*, edited by Gerard Corsane, London: Routledge, 38-53.

Hart, F.R. and Pick, J.B. 1985. *Neil M. Gunn. A Highland Life*. First published 1981. Edinburgh: Polygon.

Hayes, D. 1991. Association of Scottish visitor attractions. *Scottish Museums News*, Spring, 12-3.

Heighton, M. 1987. Museums as community enterprises. *Museums Journal*, 87(2), 61-4.

Hetherington, K. 1994. The contemporary significance of Schmalenbach's concept of the Bund, *The Sociological Review*, 42(1), 1-25.

Hewison, R. 1987. *The Heritage Industry*. London: Methuen.

Heywood, F. 2002. The lie of the land. *Museums Journal*, 102(9), 26-31.

Highland Regional Council. 1977. *Highland Regional Report*. Inverness.

Highlands and Islands Development Board. 1970? *Strath of Kildonan. Proposals for Development. Special Report 5*. Inverness: Highlands and Islands Development Board.

Highlands and Islands Development Board. 1979 (February). *The Highlands and Islands: A Contemporary Account*. Inverness: Highland and Island Development Board.

Highlands and Islands Development Board. 1981. *Wet Weather Facilities Study*. Inverness: Highland and Island Development Board.

Highlands and Islands Development Board. *Annual Reports*.

Highlands and Islands Enterprise. *Annual Reports*.

Hillier, D. 1994. Green culture. *Scottish Museums News*, 10(2), Autumn 1994, 14-6.

Hingley, R. 1993. Past, current and future preservation and management options, in *Mediaeval or Later Rural Settlement in Scotland. Management and Preservation*, edited by R. Hingley. Edinburgh: Historic Scotland, 52-61.

Hirsch, M. 1997. *Family Frames: Photography Narratives and Postmemory*, Cambridge, MA: Harvard University Press.

History Trust of South Australia. 2003. *Guiding Principles for Presenting History*. Available at: http://www.history.sa.gov.au/history/about_us/principles_history.htm [accessed 22 October 2008], 5.

Hobsbawm, E. 1980. Scottish reformers of the eighteenth century and capitalist agriculture, in *Peasants in History: Essays in Honour of Daniel Thorner*, edited by E. Hobsbawm. Oxford: Oxford University Press.

Hobsbawm, E. 1983. Introduction: Inventing tradition, in *The Invention of Tradition*, edited by E. Hobsbawm and T. Ranger, Cambridge: Cambridge University Press, 1-14.

Hodgkin, K. and Radstone S. 2003. *Contested Pasts. The Politics of Memory*. London: Routledge.

Hooper-Greenhill, E. 2000. *Museums and the Interpretation of Visual Culture*. London: Routledge.

Howe, A. 1988a. 'We have seen the land, and behold, it is very good', in *That Land of Exiles: Scots in Australia*, edited by E. Richards et al. Edinburgh: HMSO, 68-83.

Howe, A. 1988b. 'These people ... Give the prevalent tone to society', in *That Land of Exiles: Scots in Australia,* edited by E. Richards et al. Edinburgh: HMSO, 101-119.

Hunter, J. 1974-1975. The Sutherland Clearances: Sutherland in the industrial revolution. *Northern Scotland*, 2(1), 75-6.

Hunter, J. 1976. *The Making of the Crofting Community*. Edinburgh: John Donald.

Hunter, J. 1989. Address to the 1989 Annual Conference of the Scottish Crofters Union. *The Crofter*. June, 8.

Hunter, J. 1992. Time for crofters to take charge of their own destinies. *The Crofter*, June, 6.

Hunter, J. 1994. *A Dance Called America: The Scottish Highlands, the United States and Canada*. Edinburgh: Mainstream.

Hunter, J. 1995. *On the Other Side of Sorrow*. Edinburgh: Mainstream.

Hunter, J. 1996. *Glencoe and the Indians: A Real-Life Family Saga which Spans Two Continents, Several Centuries and More than Thirty Generations to Link Scotland's Clans with the Native Peoples of the American West*. Edinburgh: Mainstream.

Hunter, J. 1997. How should we treat America's growing obsession with its Scottish connections? *Scotland on Sunday*, 11 May, 17.

Hunter, J. 2006. *Gàidhealtachd Alba: Tìr fo dheasbad* – The Scottish Highlands: A contested country, in *Fonn's Duthchas – Land and Legacy*, commissioned by the National Museums of Scotland, the National Library of Scotland, the National Galleries of Scotland, Highland and Island Museums and the Scottish Executive. Edinburgh: National Museums of Scotland, 2-59.

Hunter, M. 1988. *A Pistol in Greenyards*. First published 1965. Edinburgh: Canongate Publishing.

Hutchinson, R. 2005. Fifty years on, and McLean still captivates his young readers. *West Highland Free Press*, 25 March.

Hutchison, M. 2009. Dimensions for a folding exhibition. Exhibiting diversity in theory and practice in the *Migration Memories* exhibitions. *Humanities Research* [online], 15(2), 67-92. Available at: http://epress.anu.edu.au/hrj2009_02_citation.html [accessed 12 December 2009].

Hutton, P.H. 1993. *History as an Art of Memory*. Hanover: University Press of New England.

Huyssen, A. 1995. *Twilight Memories. Marking Time in a Culture of Amnesia*. London: Routledge.

Inverness Courier. 1930. Highland exhibition. Opening ceremony by Sir D.Y. Cameron. Proposed folk museum, 5 August.

Jameson, E. and Mouat, J. 2006. Telling differences: The forty-ninth parallel and historiographies of the west and nation. *Pacific Historical Review*, 75 (2), 183-230.

Jenkins, J.G. 1989. The collection of material objects and their interpretation, in *Museum Studies in Material Culture*, edited by S.M. Pearce. Leicester: Leicester University Press, 119-24.

Jenkinson, P. 1989. Material culture, people's history and populism: Where do we go from here?, in *Museum Studies in Material Culture*, edited by S.M. Pearce. Leicester: Leicester University Press, 139-52.

Johnston, T. 1946. *The History of the Working Classes in Scotland.* 4th edition. Glasgow: Unity Publishing Company.

Johnston, T. 1999. *Our Scots Noble Families.* First published 1909. Glendaruel: Argyll Publishing.

Joly, M.-H. and Compère-Morel, T. 1998. *Des Musées d'histoire pour l'avenir.* Paris: Noésis.

Joly, M.-H., 1998. Les musées d'histoire, in *Des musées d'histoire pour l'avenir*, edited by M.-H. Joly and T. Compère-Morel. Paris: Noésis, 57-86.

Jones, M. 1995. From Haggis to Home Rule. *Museums Journal*, 95(2), 26-7.

Jones, M. 2000. Why a Museum of Scotland? Aspirations and expectations, in *Heritage and Museums. Shaping National Identity*, edited by J.M. Fladmark. Shaftesbury: Donhead Publishing, 3-16.

Jordanova, L. 1989. Objects of knowledge: A historical perspective on museums, in *The New Museology*, edited by P. Vergo. London: Reaktion Books, 22-40.

Karlin, G.J. 1904. The pavilion system in museums. *Museums Journal*, 3(12), 386-96.

Karp, I. 1992. Introduction. Museums and communities: The politics of public culture, in *Museums and Communities: The Politics of Public Culture*, edited by I. Karp et al. Washington, DC: Smithsonian Institution Press.

Karp, I., Kreamer C.M. and Levine S.D. 1992. *Museums and Communities: The Politics of Public Culture.* Washington, DC: Smithsonian Institution Press.

Kavanagh, G. 1989. Objects as evidence, or not? in *Museums Studies in Material Culture*, edited by S.M. Pearce. Leicester: Leicester University Press, 125-37.

Kavanagh, G. 1990. *History Curatorship.* Leicester: Leicester University Press.

Kavanagh, G. 1991. Mangles, muck and myths: Rural history in Britain. *Rural History*, 2(2), 187-203.

Kennedy, M. 1999. 'Lochaber no more': A critical examination of emigration mythology, in *Myth, Migration and the Making of Memory: Scotia and Nova Scotia c. 1700-1990*, edited by M. Harper and M.E. Vance. Edinburgh: John Donald and Halifax: Fernwood Publishing, 267-297.

Kershaw, B. 1992. *The Politics of Performance. Radical Theatre as Cultural Intervention.* London: Routledge.

Kidd, C. 1995. Teutonist ethnology and Scottish nationalist inhibition 1780-1880. *Scottish Historical Review*, 197, 45-68.

Kinchin, P. and Kinchin, J. 1988. *Glasgow's Great Exhibitions: 1888, 1901, 1911, 1938, 1988*. Wendlebury: White Cockade Publishing.

Kirmayer, L.J. 1996. Landscapes of memory: Trauma, narrative, and dissociation, in *Tense Past. Cultural Essays in Trauma and Memory*, edited by P. Antze and M. Lambek. London: Routledge, 173-98.

Kirshenblatt-Gimblett, B. 1998. *Destination Culture: Tourism, Museums and Heritage*. Berkeley: University of California Press.

Knox, J.A. 1785. *A View of the British Empire, More Especially Scotland; With Some Proposals for the Improvement of that Country, the Extension of its Fisheries and the Relief of the People*. London.

Knox, W. 1984. *Scottish Labour Leaders 1918-1939*. Edinburgh: Mainstream.

Laborie, P. 2001. *Les Français des années troubles*. Paris: Desclée de Brouwer.

Land Use Working Group of the Highland Green Party. 1989. *A Rural Manifesto for the Highlands: Creating the Second Great Wood of Caledon*. Inverness: Highland Printers.

Latham, P.R. 1883. The deterioration of mountain pastures, and suggestions for their improvements. *Transactions of the Royal Highland and Agricultural Society of Scotland*, 15, 111-30.

Le Goff, J. 1998. Introduction des Entretiens du Patrimoine, in *Patrimoines et passions identitaires. Actes des entretiens du patrimoine 1997*, edited by J. Le Goff. Paris: Fayard, 9-13.

Linklater, M. 1998. Timely awakening of our heritage. *Scotland on Sunday*, 29 November, 15.

Livingston, W. 1995. A message for the poet, in *Tuath is Tighearna. Tenants and Landlords*, edited by D.E. Meek. Edinburgh: Scottish Academic Press, 199-20.

Loch, J. 1820. *An Account of the Improvements on the Estates of the Marquess of Stafford*. London.

Lochaber News. 1999. Centre's focus on history, 3 July, 8.

Lorente, P. 1995. The city's beating art. *Museums Journal*, 95(10), 29-30.

Lowenthal, D. 1985. *The Past is a Foreign Country*. Cambridge: Cambridge University Press.

Lowenthal, D. 1996. *The Heritage Crusade and the Spoils of History*. Harmondsworth: Penguin Books.

Lumley, R. 1987. Museums in a post-modern world. *Museums Journal*, 87(2), 81-3.

Lumley, R. 1988. *The Museum Time Machine: Putting Cultures on Display*. London: Routledge.

Lumley, R. 2005. The debate on heritage reviewed, in *Heritage, Museums and Galleries: An Introductory Reader*, edited by G. Corsane. London: Routledge, 15-25.

Lynch, M. 1992. *Scotland: A New History*. London: Pimlico.

Lynch, M. 2001. *The Oxford Companion to Scottish History*. Oxford: Oxford University Press.

MacAskill, J. 1999. *We Have Won the Land*. Stornoway: Acair.
MacColl, A. W. 2006. *Land, Faith and the Crofting Community: Christianity and Social Criticism in the Highlands of Scotland 1843-1893*. Edinburgh: Edinburgh University Press.
MacColla, F. 1975. *Ro Fhada Mar So A Tha Mi. Too Long in this Condition*. Thurso: Caithness Books.
MacColla, F. 1983. *And the Cock Crew*. First published 1945. London: Souvenir Press.
MacCuish, D.J. 1976-1978. Ninety years of crofting legislation and administration. *Transactions of the Gaelic Society of Inverness*, vol. L, 296-326.
Macdiarmid, J.M. 1926. *The Deer Forests and How They Are Bleeding Scotland White*. 2nd edition. Glasgow: Scottish Home Rule Association.
Macdonald P. and Macdonald A. 2009. Marginal lands? An overview of the environmental contexts of cultural landscapes in the Highlands and Islands. *International Journal of Heritage Studies*, 15 (2-3), 108-141.
Macdonald, J. 1877. On the agriculture of the Counties of Ross and Cromarty. *Transactions of the Royal Highland and Agricultural Society of Scotland*, 9, 67-209.
Macdonald, J. 1880. On the agriculture of the County of Sutherland. *Transactions of the Royal Highland and Agricultural Society of Scotland*, 12, 1-90.
MacDonald, J. n.d. *A Short History of Crofting in Skye*. Skye Museum of Island Life. Was available at: http://pages.eidosnet.co.uk/~skye/crofting.html [accessed 10 June 2008].
Macdonald, S. 1997. A people's story. Heritage, identity and authenticity, in *Touring Cultures. Transformations of Travel and Theory*, edited by C. Rojek and J. Urry. London: Routledge, 155-75.
Macdonald, S. 2003. Museums, national, postnational and transcultural identities. *Museum and Society*, 1(1), 1-16.
Macdonald, S. 2006. *A Companion to Museum Studies*. Oxford: Blackwell Publishing.
Macdonald, S. 2007. On 'old things': The fetishization of past everyday life, in *Cultural Heritage. Critical Concepts in Media and Cultural Studies*, edited by L. Smith. London: Routledge, 269-86.
Macdonald, S. and Fyfe, G. 1996. *Theorising Museums: Representing Identity and Diversity in a Changing World*. Oxford: Blackwell Publishing.
MacEwen, A. 1930. Correspondence. *Inverness Courier*, 9 September.
Macinnes, A.I. 1988. Scottish Gaeldom: The first phase of Clearance, in *People and Society*, edited by T.M. Devine and R. Mitchison. Edinburgh: John Donald, 70-90.
Macinnes, A.I. 1996. *Clanship, Commerce and the House of Stuart 1603-1788*. East Linton: Tuckwell Press.
Macintyre, S. 2004. *A Concise History of Australia*. First published 1999. Cambridge: Cambridge University Press.

Macintyre, S. and Clark, A. 2003. *The History Wars*. Melbourne: Melbourne University Publishing.
Mackay, D. 1993. Scottish rural settlement: Preserving a people's past, in *Mediaeval or Later Rural Settlement in Scotland. Management and Preservation*, edited by R. Hingley. Edinburgh: Historic Scotland, 43-51.
Mackenzie, A. 1986. *The History of the Highland Clearances*. First published 1883. Perth: Melven Press.
Mackenzie, A.F.D. 2004. Re-imagining the land, North Sutherland, Scotland. *Journal of Rural Studies*, 20, 273-287.
MacLachlan, J. 1995. Climbing up towards Ben Shiant, in *Tuath is Tighearna. Tenants and Landlords*, edited by D.E. Meek. Edinburgh: Scottish Academic Press, 192-93.
Maclauchlan, T. 1849. *The Depopulation System in the Highlands*. Edinburgh: Johnstone and Hunter.
MacLean, J. 1995a. Manitoba, in *Tuath is Tighearna. Tenants and Landlords*, edited by D.E. Meek. Edinburgh: Scottish Academic Press, 205-6.
MacLean, J. 1995b. The Queen's Commissioners, in *Tuath is Tighearna. Tenants and Landlords*, edited by D.E. Meek. Edinburgh: Scottish Academic Press, 238-9.
MacLean, N. 1953. *The Former Days*. First published 1945. London: Hodder and Stoughton.
MacLennan, H.D. 2001. *Gu fearann an òir*: To the land of gold, in *Scatterlings of Empire*, edited by W. Prest and G. Tulloch. Special issue of *Journal of Australian Studies*, 68, 2001, 44-53.
MacLeod, D. 1965. Western round-up. Story of Skye told in 'old blackhouse'. *Press and Journal*, 25 May, 6.
MacLeod, D. 1976. History creates jobs for future. *Northern Times*, 16 January, 8.
MacLeod, D. 1994. *Gloomy Memories in the Highlands of Scotland*. Facsimile reprint of 1892 edition. New Delhi: B. Jain Publishers.
MacLeod, M. 1995. Song to the people of Bernera, in *Tuath is Tighearna. Tenants and Landlords*, edited by D.E. Meek. Edinburgh: Scottish Academic Press, 210-12.
Macmillan, D. 1992. *Symbols of Survival. The Art of Will Maclean*. Edinburgh: Mainstream Publishing.
MacPherson, I. 1933. *Land of our Fathers*. London: Jonathan Cape.
MacPherson, J.A. 1995. Beyond the memories: Drawing strength from the diaspora, in *Sharing the Earth: Local Identity in Global Culture*, edited by J.M. Fladmark. London: Donhead Publishing, 279-288.
MacPherson, M. 1972. *The Battle of the Braes*. London: Collins.
MacPherson, M. 1985. Crofters and the Crofters Commission, in *The Land for the People*, edited by Irene Evans and Joy Hendry. Blackford: Scottish Socialist Party, 19-21.
Marr, A. 1992. *The Battle for Scotland*. Harmondsworth: Penguin.

Martin, J.-C. 2000. Histoire, mémoire et oubli. Pour un autre régime d'historicité. *Revue d'histoire moderne et contemporaine*, 47(4), October-December, 783-804.

Mather, A.S. 1985. The rise and fall of government-assisted land settlement in Scotland. *Land Use Policy*, 2(3), 217-24.

McCann, A. 1986. Caithness museums service. *Scottish Museums News*, Autumn, 19.

McCrone, D. 1992. *Understanding Scotland. The Sociology of a Stateless Nation.* London: Routledge.

McCrone, D. 1997. *Land, Democracy and Culture in Scotland: 4th John McEwen Memorial Lecture*, 10 Oct 1997. Available at: http://www.caledonia.org.uk/land/mccrone.htm [accessed 18 June 2007].

McCrone, D., Morris A. and Kiely R. 1995. *Scotland – the Brand: The Making of Scottish Heritage*. Edinburgh: Edinburgh University Press.

McGeachy, R.A.A. 2005. *Argyll 1730-1850: Commerce, Community and Culture.* Edinburgh: John Donald.

McGrath, J. 1981a. *The Cheviot, the Stag and the Black, Black Oil.* First published 1974. London: Methuen.

McGrath, J. 1981b. *A Good Night Out.* London: Eyre Methuen.

McIntyre, D. and Wehner K. 2001. *National Museums. Negotiating Histories.* Canberra: National Museum of Australia.

McKean, C. 2000. *The Making of the Museum of Scotland.* Edinburgh: National Museums of Scotland Publishing.

McLean, A.C. 1985. *A Sound of Trumpets.* First published 1967. Edinburgh: Canongate Publishing.

McLean, A.C. 1989. *Ribbon of Fire.* First published 1962. Edinburgh: Canongate Publishing.

McLean, F. and Cooke, S. 2003. The National Museum of Scotland: A symbol for a new Scotland? *Scottish Affairs*, 45, 111-127.

McShane, I. 2001. Challenging or conventional? Migration history in Australian museums, in *National Museums. Negotiating Histories*, edited by D. McIntyre and K. Wehner. Canberra: National Museum of Australia, 122-133.

Medhurst, B. 1999. *Scottish Reference Book. Acknowledging the Scottish Immigrants and their Descendants, their Achievements and Legacy to the State of South Australia.* The Scottish Association of South Australia.

Meek, D.E. 1995. *Tuath is Tighearna. Tenants and Landlords.* Edinburgh: Scottish Academic Press.

Meek, D.E. 1998. Gaelic bible, revival and mission: The spiritual rebirth of the nineteenth-century Highlands, in *The Church in the Highlands*, edited by J. Kirk. Edinburgh: Scottish Church History Society.

Menabney, J. 2001. The 'Baile Failte' project: Displacement of language and social evolution in the Scottish Highlands. *Museum International*, 53(3), 38-39.

Message, K. 2006. *New Museums and the Making of Culture.* Oxford: Berg.

Message, K. 2009. Culture, citizenship and Australian multiculturalism. The contest over identity formation at the National Museum of Australia. *Humanities Research* [Online], 15(2), 23-48. Available at: http://epress.anu.edu.au/hrj2009_02_citation.html [accessed 12 December 2009].

Middleton, D. and Edwards, D. 1990. *Collective Remembering*. London: Sage.

Migration Museum. 2003. *Background on the Museum* [Online: History Trust of South Australia]. Available at: http://www.sa.gov.au/migration/about/history.htm [accessed 21 October 2008].

Milton, N. 1978. *John Maclean. In the Rapids of Revolution. Essays, Articles and Letters 1902-23*. London: Allison and Busby.

Misztal, B. 2003. *Theories of Social Remembering*. Maidenhead: Open University Press.

Mitchell, H. 1880. *The Past in the Present: What is Civilization*. Edinburgh: David Douglas.

Morrison, W. 1995. Ullapool museum set to open. *Press and Journal*, 13 December.

Mortimer Wheeler, R.E. 1934. Folk museums. *Museums Journal*, 34(5), 191-96.

Munro, N. 1911. Introduction. *Leabhar a'Chlachain. Home Life of the Highlanders 1400-1746*. Glasgow: Printed for the Highland Village Association by R. Maclehose, v-vi.

Museums and Galleries Commission. 1986. *Museums in Scotland*. London: HMSO.

Museums Journal. 1907. The Dundee Conference, 7(2), 41-52.

Museums Journal. 1930. Open-air folk museums, 29(11), 377-81.

Museums Journal. 1931. Folk museum committee, 30(8), 300.

Museums Journal. 1945. News from the museums, 45(7), 118.

Museums Journal. 1946. St Fagans Castle. A folk museum for the Welsh nation, 46(6), 104-6.

Museums Journal. 1972. Session 2: Museums and their responsibility for environmental conservation, 72(3), 93-100.

Museums Journal. 1995. News in brief, 95(2), 8.

Museums Journal. 1998. Funds exit from Scotland, 98(5), 7.

Nederveen Pieterse, J. 2005. Multiculturalism and museums: Discourse about others in the age of globalization, in *Heritage, Museums and Gallerie. An Introductory Reader*, edited by G. Corsane. London: Routledge, 163-83.

New Statistical Account of Scotland (15 vols.). 1834-1845. Edinburgh.

Newman, A. 2001. Social exclusion zone. *Museums Journal*, 101(9), 34-36.

Noble, R. 1977-1978. The changing role of the Highland Folk Museum. *The Aberdeen University Review*, xlvii, 142-47.

Nora, P. 1997a. *Les lieux de mémoire* (3 vols.). Paris: Editions 'Quarto' Gallimard.

Nora, P. 1997b. Présentation, in *Les lieux de mémoire* (3 vols.), edited by P. Nora. Paris: Editions 'Quarto' Gallimard, 15-22.

Northern Times. 1976a. Planners want to give 'untidy' Helmsdale priority facelift, 29 February, 4.

Northern Times. 1976b. Farr museum opened, 9 July, 7.
Northern Times. 1987a. Helmsdale's showpiece centre, 3 July, 1.
Northern Times. 1987b. Back to the future in Helmsdale, 3 July, 8-9.
Northern Times. 1987c. Community news: 'history in the making', 24 July, 2.
Northern Times. 1988a. Major attraction or 'damn disgrace', 5 August, 1.
Northern Times. 1988b. Happy mix of glamour and history, 26 August, 1.
Northern Times. 1988c. The Cartland Saga, 2 September, 1.
Northern Times. 2008. Heritage centre 'not what it used to be', 25 July, 1.
O'Neill, J. 1993. *So Far from Skye*. First published 1992. London: Puffin Books.
O'Neill, M. 1987. Quantity vs quality or what is a community museum anyway? *Scottish Museums News*, Spring, 5-7.
O'Neill, M. 1990. People as a driving force. *Museums Journal*, 90(2), 21.
Ogilvy, G. 1998. Lairds ready to apologise for Highland Clearances. *Scotland on Sunday*, 26 April, 3.
Paton, D. 2006. *The Clergy and the Clearances. The Church and the Highland Crisis 1790-1850*. Edinburgh: John Donald.
Peate, I. 1934. A folk museum for Wales. *Museums Journal*, 34(6), 229-31.
Peate, I. 1941. The place of folk culture in the museum. *Museums Journal*, 41(3), 45-50.
Peate, I. 1947. Swedish Folk Museums. *Museums Journal*, 46(10), 201-2.
Peate, I. 1949. Lectures on Museums II: The Folk Museum. *Museums Journal*, 49(3), 67-70.
Peate, I. 1954. The display of folk material. *Museums Journal*, 54(1), 4-5.
Pedersen, R. 1995. Scots Gaelic as a tourism asset, in *Sharing the Earth: Local Identity in Global Culture*, edited by J.M. Fladmark. London: Donhead Publishing, 289-98.
Phillips, M.S. 2004. Distance and historical representation. *History Workshop Journal*, 57, 123-41.
Pigott, P.H. 1975. *Museums in Australia 1975: Report of the Committee of Inquiry on Museums and National Collections Including the Report of the Planning Committee on the Gallery of Aboriginal Australia*. Canberra: Australian Government Publishing Service.
Pomian, K. 1998. De l'histoire, partie de la mémoire, à la mémoire, objet d'histoire. *Revue de métaphysique et de morale*, 1 (Jan-March), 63-110.
Popular Memory Group. 1982. Popular memory: Theory, politics, methods, in *Making Histories: Studies in History-Writing and Politics*, edited by R. Johnson et al., Centre for Contemporary Cultural Studies. London: Hutchison Education, 205-52.
Prebble, J. 1969. *The Highland Clearances*. First published 1963. Harmondsworth: Penguin Books.
Prebble, J. 1988. *The King's Jaunt*. London: Collins.
Prebble, J. 1993. *Landscape and Memories: An Intermittent Autobiography*. London: HarperCollins.
Press and Journal. 1965. Skye cottage transformed into museum, 12 May.

Press and Journal. 1978. Luib cottage marks another step in Skye plan to rebuild the past, 2 March.

Press and Journal. 1997. Independent museums fear for their future, 10 November.

Press and Journal. 1998. 'Fair deal' call for museums, 3 March.

Rattue, J. 1996. History and folklore, in *Making Histories in Museums*, edited by G. Kavanagh. London: Leicester University Press, 216-27.

Report of Her Majesty's Commissioners of Inquiry into the Conditions of the Crofters and Cottars in the Highlands and Islands of Scotland. Vol. 1. 1884. Edinburgh.

Report of the Commission of Enquiry into Crofting Conditions. 1954. Edinburgh: HMSO.

Report of the Committee on the National Museum of Antiquities of Scotland. 1952 (Cmd 8604), HMSO.

Richards, E. 1973a. *The Leviathan of Wealth: The Sutherland Fortune in the Industrial Revolution*. London: Routledge.

Richards, E. 1973b. How tame were the Highlanders during the Clearances? *Scottish Studies*, 17, 35-50.

Richards, E. 1974. Patterns of Highland discontent, 1790-1860. *Popular Protest and Public Order. Six Studies in British History, 1790-1920*, edited by R. Quinault and J. Stevenson. London: Allen and Unwin, 75-114.

Richards, E. 1974-5. The Sutherland Clearances: New evidence from Dunrobin. *Northern Scotland*, 2(1), 59-74.

Richards, E. 1982. *A History of the Highland Clearances: Agrarian Transformation and the Evictions 1746-1886*. Vol. 1. London: Croom Helm.

Richards, E. 1982-3. Highland emigrants to South Australia in the 1850s. *Northern Scotland*, 5, 1-29.

Richards, E. 1985. *A History of the Highland Clearances: Emigration, Protest, Reasons*. Vol. 2. London: Croom Helm.

Richards, E. 1988. Scottish Australia 1788-1914, in *That Land of Exiles: Scots in Australia*, edited by E. Richards et al. Edinburgh: HMSO, 8-39.

Richards, E. 1993. How did poor people emigrate from the British Isles to Australia in the nineteenth century? *Journal of British Studies*, 32(3), 250-279.

Richards, E. 1999. *Patrick Sellar and the Highland Clearances: Homicide, Eviction and the Price of Progress*. Edinburgh: Polygon.

Richards, E. 2000. *The Highland Clearances: People, Landlords and Rural Turmoil*. Edinburgh: Edinburgh University Press.

Richards, E. 2001. Highland and Gaelic Immigrants, in *The Australian People. An Encyclopedia of the Nation, its People and their Origins*, edited by James Jupp. Cambridge: Cambridge University Press, 649-655.

Richards, E. 2007. *Debating the Highland Clearances*. Edinburgh: Edinburgh University Press.

Richardson, D. 1999. *The Curse on Patrick Sellar. An Incident in Sutherland in 1814*. Stockbridge: Longstock Books.

Ricoeur, P. 1998a. La marque du passé. *Revue de métaphysique et de la morale*, 1 (Jan-March), 7-31.
Ricoeur, P. 1998b. Vulnérabilité de la mémoire, in *Patrimoine et passions identitaires*, edited by J. Le Goff. Paris: Fayard, 17-31.
Ricoeur, P. 2000. *La mémoire, l'histoire, l'oubli*. Paris: Seuil.
Robin, R. 2003. *La mémoire saturée*. Paris: Stock.
Ross, D. 1852. *The Scottish Highlanders: Their Present Sufferings and Future Prospects*. Glasgow.
Ross, D. 1853. *The Glengarry Evictions; or Scenes at Knoydart, in Inverness-shire*. Glasgow.
Ross, D. 1854a. *The Russians of Ross-shire, or Massacre of the Rosses in Strathcarron, Ross-shire*. Glasgow.
Ross, D. 1854b. *Real Scottish Grievances*. Glasgow.
Ross, J. 2005. 'Lost' village resurrected for future generations. *The Scotsman*, 16 August.
Rousso, H. 1991. *The Vichy Syndrome: History and Memory in France since 1944*. Cambridge, MA: Harvard University Press.
Rousso, H. 1993a. Une justice impossible. L'épuration et la politique antijuive de Vichy. *Annales ESC*, 3 (May-June), 745-770.
Rousso, H. 1993b. Le souvenir de la seconde guerre mondiale et des persécutions antisémites. Contexte historique et repères chronologiques. *Annales ESC*, 3 (May-June), 799-809.
Rousso, H. 2003. Justice, history and memory in France: Reflections on the Papon trial, in *Politics of the Past. On Repairing Historical Injustices*, edited by J. Torpey. Lanham. Maryland: Rowman and Littlefield, 277-293.
Royal Commission on National Museums and Galleries. 1929. *Final Report, Part 1. General Conclusions and Recommendations* (Cmd 3401), 20 September 1929.
Ryder, J. 1995. Registration phase 2 launch – 17 July 1995. *Scottish Museums News*, 11(2), Autumn, 2-3.
Ryder, J. 1996. Lottery funding: The first year's experience. *Scottish Museums News*, 12(1), Spring, 8-10.
Samuel, R. 1996. *Theatres of Memory*. London: Verso.
Samuel, R. and Thompson, P. 1990. *The Myths we Live by*. London: Routledge.
Sandell, R. 1999. The regeneration game. *Museums Journal*, 99(7), 30-31.
Schildkrout E. 2004. Ambiguous messages and Ironic Twists: *Into the Heart of Africa* and *The Other Musem*, in *Museum Studies: An Anthology of Contexts*, edited by B. Messias Carbonell. Oxford: Blackwell, 181-192.
Schlereth, T.J. 2004. Collecting ideas and artifacts. Common problems of history museums and history texts, in *Museum Studies: An Anthology of Contexts*, edited by B.M. Carbonell. Oxford: Blackwell, 335-47.
Schwartz, B. 1982. The social context of commemoration: A study in collective memory. *Social Forces*, 61(2), 374-402.
Scotland on Sunday. 1998. Letters to the editor, 3 May, 16.

Scots Magazine. 1934. Books and other things, 22(2), 149.
Scottish Arts Council, Scottish Museums Council, Scottish Film Council and Convention of Scottish Local Authorities. 1993. *Charter for the Arts*. Edinburgh: HMSO.
Scottish Crofters Union and Royal Society for the Protection of Birds. 1992. *Crofting and the Environment: A New Approach*. Inverness: Highland Printers.
Scottish Daily Record. 1997. Firm favourite, 24 September.
Scottish Executive. 1999. *Celebrating Scotland. A National Cultural Strategy*. Available at: http://www.scotland.gov.uk/nationalculturalstrategy/docs/csnse-00.htm.
Scottish Executive. 2000. *Creating our Future ... Minding our Past. Scotland's National Cultural Strategy*. Edinburgh: Scottish Executive.
Scottish Exhibition of History, Art and Industry. May to October 1911. Official Catalogue (Industrial Section). Kelvingrove Park. 1911. Glasgow.
Scottish Exhibition of National History, Art and Industry. Palace of History. Catalogue of Exhibits. Vol. 1. 1911. Glasgow.
Scottish Farmer. 1978. Historic church saved to tell an old story, 24 June.
Scottish Museums Council, National Museums of Scotland and National Galleries of Scotland. 1999. *A National Strategy for Scotland's Museums*.
Scottish Museums Council. 1988-1989. *Annual Reports and Accounts*.
Scottish Museums Council. 1990. *Scottish Museums and Galleries: The Guide*. Aberdeen: Aberdeen University Press.
Scottish Museums Council. 1998. Best value for museums: A corporate approach. *Museum Issues*, 8.
Scottish Museums Council. 2001. Highland Independent Museums Assessment Summary. *Museum Issues*, 11.
Scottish Museums Council. 2002. *A Collective Insight. Scotland's National Audit*.
Scottish Museums News. 1990. Scottish museum of the year awards, Autumn 1990, 10-11.
Scottish Office. 1997. *Scotland's Parliament* (Cm 3658), HMSO.
Scottish Parliament. 2000. *Official Report*, 8(7), 27 September, Col. 700 and 706.
Scottish Tourist Board. 1998. *Scottish Visitor Attractions: Review 1997*. Edinburgh: Scottish Tourist Board.
Sebastian, P. 2007. Mobilizing communities and sharing stories: The role of the Immigration Museum in one of the most culturally diverse cities in the world. *Museum International*, 59(1/2), 151-159.
Sherman, D.J. and Rogoff, I. 1994. *Museum Culture. Histories, Discourses, Spectacles*. London: Routledge.
Silverstone, R. 1989. Heritage as media: Some implications for research, in *Heritage Interpretation*, vol. 2, edited by D.L. Uzzell. London: Belhaven Press, 138-48.
Simpson, M. 1992. Celebration, commemoration or condemnation? *Museums Journal*, 92(3), 28-31.

Small, G. 2002. *The History of the Isle of Arran Heritage Museum*, unpublished paper.
Smith, I.C. 1981. *Consider the Lilies*. First published 1968. Exeter: A. Wheaton.
Smith, L. 2006. *Uses of Heritage*. London: Routledge.
Smout, T.C. 1983. Tours in the Scottish Highlands from the eighteenth to the twentieth centuries. *Northern Scotland*, 5, 99-121.
Smout, T.C. 1986. *A Century of the Scottish People 1830-1950*. London: Fontana Press.
Smout, T.C. 1987. *A History of the Scottish People 1560-1830*. First published 1969. London: Fontana Press.
Smout, T.C. 1989. Problems of nationalism, identity and improvement in later eighteenth-century Scotland, in *Improvement and Enlightenment: Proceedings of the Scottish Historical Studies Seminar*, edited by T.M. Devine. Edinburgh: John Donald, 1-21.
Smout, T.C. 1993. *The Highlands and the Roots of the Green Consciousness 1750-1990*. Occasional Paper No. 1, Battleby: Scottish Natural Heritage.
Smout, T.C. 1994. Perspectives on the Scottish identity. *Scottish Affairs*, 6, 1994, 101-113.
Sprott, G. 2000. Jock Tamson's bairns. Ethnicity and identity, in *Heritage and Museums. Shaping National Identity*, edited by J.M. Fladmark. Shaftesbury: Donhead Publishing, 103-114.
Statistical Account of Scotland (21 vols). 1791-1799. Edinburgh.
Stephens, M. 2006. Peate, Iorwerth Cyfeiliog (1901-1982). *Oxford Dictionary of National Biography*. 2004. Oxford: Oxford University Press. Online edition, October 2006, Available at: http://www.oxforddnb.com/view/article/55462 [accessed 1 September 2009].
Stewart, D. 1822. *Sketches of the Character, Manners, and Present State of the Highlanders of Scotland*. Vol. 1. Edinburgh: Longman et al.
Stora, B. 2007. *La guerre des mémoires. La France face à son passé colonial*. La Tour d'Aigues: Aube poche.
Strathnaver Museum Trustees. n.d. *Draft Proposal for Strathnaver Museum*.
Strathnaver Museum. n.d. *Dr Ian Grimble: Writer/Historian. Radio and TV Producer. Presenter. Co-Founder of Strathnaver Museum. 1921-1995*. Strathnaver Museum.
Sutherland, I. 1996. Heritage is no quick fix, *The Herald*, 12 October, 14.
Sutherland, S.R. 1974-75. The Sutherland Clearances: Ethics and economics in the Sutherland clearances. *Northern Scotland*, 2(1), 77-83.
Szekeres, V. 1992. Whose cultural values? Conserving relics from major historical events: Migration and multiculturalism. Paper to Australia ICOMOS conference: Whose Cultural Values? Sydney, 19-20 November 1992.
Szekeres, V. 2005. *Changing Headsets: The Impact of Museums on Social Thinking: 2005 Australia-Israel Hawke Lecture*, 9 March 2005. Available at: http://www.unisa.edu.au/hawkecentre/events/2005events/2005ai_lecture_Szekeres.asp [accessed 21 November 2008].

Taylor, M. 2004. What is in a 'national' museum? The challenge of collecting policies at the National Museums of Scotland, in *Museums and the Future of Collecting*, edited by S. Knell. First published 1999. Aldershot: Ashgate, 165-78.

Taylor, T.M. 1954. Letter to N.M. Gunn. National Library of Scotland: Dep. 209, Box 16, folder 1, June 1954.

The Clarion of Skye. 1955. District news: Glendale, 59 (December), 14.

The Clarion of Skye. 1956. Restoring the crofter's confidence, 70 (November), 1.

The Crofter. 1992. Time for crofters to take charge of their own destinies, 24 (June), 6.

The Economist. Past and present in Scotland, 28 November 1998, 42.

The Guardian. 1998. Arts – You take the high road, 4 April, 6.

The Herald. 1998. Land reformers unmoved by plan for Clearances apology, 27 April, 9.

The Scotsman. 1945. Books of the Day – New Novels, 13 December, 7.

The Scottish Banner. 1999. Unearthing the lost village, September, 14.

Thelen, D. 1989. Memory and American History? *Journal of American History*, 75(4), 1117-1129.

Thompson. P. 1988. *The Voice of the Past*. 2nd edition. Oxford: Oxford University Press.

Timespan. 2008. *Newsletter*, 1.

Todorov, T. 1995. *Les abus de la mémoire*. Paris: Arléa.

Todorov, T. 2001. La vocation de la mémoire. *Cahiers français*, 303 (July-August), 3-7.

Tonkin, E. 1992. *Narrating our Pasts. The Social Construction of Oral History*. Cambridge: Cambridge University Press.

Touchstone Heritage Management Consultants. 1997. *Rosal Clearance Village. Interpretive Panels. Final Texts* (October 1997).

Tranter, N. 1993. *The Gilded Fleece*. First published 1942. Edinburgh: B&W Publishing.

Tyrrell, A. 2003. What shall we do with the Duke? Public memory and culture wars in the Scottish style. *Humanities Research*, 10(2), 21-28.

UNESCO. 1982. Mexico City Declaration on Cultural Policies, adopted by the World Conference on Cultural Policies, Mexico, 6 August 1982, in *World Conference on Cultural Properties, Mexico City, 26 July-6 August 1982, Final Report*. Paris: UNESCO, 41-6.

Urry, J. 1996. How societies remember the past, in *Theorising Museums. Representing Identity and Diversity in a Changing World*, edited by S. Macdonald and G. Fyfe. Oxford: Blackwell, 45-65.

Uzzell, D. 1998. Interpreting our heritage: A theoretical interpretation, in *Contemporary Issues in Heritage and Environmental Interpretation. Problems and Prospects*, edited by D. Uzzell and R. Ballantyne. London: Stationary Office, 11-25.

Valensi, L. 1993. Présence du passé, lenteur de l'histoire. *Annales ESC*, 3, May-June, 491-500.
Varine, H. de. 2005a. *New Museology and the De-Europeanisation of Museology*. Paper to the Guizhou International Forum of Ecomuseums on Communication and Exploration, Guyiang, Guizhou, China, 1-6 June 2005, available at: http://www.trentinocultura.net/doc/soggetti/ecomusei/Atti_Convegno_eng.pdf [accessed 7 June 2008].
Varine, H. de. 2005b. *Les racines du futur. Le patrimoine au service du développement local*. Lusigny-sur-Ouche: Asdic Editions.
Vergo, P. 1989. *The New Museology*. London: Reaktion Books.
Vize, R. 1999. Good, better, best. *Museums Journal*, 99(8), 39-43.
Walsh, K. 1992. *The Representation of the Past: Museums and Heritage in the Post-Modern World*. London: Routledge.
Ward, A. 1996. Scotland: Plaid to the Bone. *National Geographic Magazine*, 90(3), September, 2-27.
Warren, C. 2002. *Managing Scotland's Environment*. Edinburgh: Edinburgh University Press.
Watson, G. 1990. Ross and Cromarty district council museums service. *Scottish Museums News*, Autumn 1990, 17-9.
Watson, G. 1995. Museums from the ground up. A community approach to development, in *Sharing the Earth*, edited by J.M. Fladmark. London: Donhead Publishing, 253-264.
Watson, S. 2007a. Museums and their communities, in *Museums and their Communities*, edited by S. Watson. London: Routledge, 1-23.
Watson, S. 2007b. *Museums and their Communities*. London: Routledge.
Weil, S. 2007. The museum and the public, in *Museums and their Communities*, edited by S. Watson. London: Routledge, 32-46.
Weir, T. 1972. My month: A "lost wood" in Skye. *The Scots Magazine*, 97(6), September, 546-551.
West Highland Free Press. 1976. Eleventh hour bid against croft reform, 6 February, 1.
West Highland Free Press. 1981. Know your Clearances, 10 April.
West Highland Free Press. 1993a. Skye's new high-tech visitor centre open next week, 5 March, 11.
West Highland Free Press. 1993b. Public funds 'should not have been used for heritage centre', 12 March, 3.
West Highland Free Press. 1995. Local groups fear for lottery fund bids, 8 December, 18.
West Highland Free Press. 1996a. *Aros* plan opposed, 12 January, 1.
West Highland Free Press. 1996b. Controversial Skye heritage centre plan recommended for go-ahead, 26 January, 16.
West Highland Free Press. 1996c. Another £500,000 for Portree cultural centre, 19 July, 5.

Westbrook S., Callaghan S. and Burns, B. 1999. *Assessment of Highland Independent Museums*. Final Report for the Scottish Museums Council, the Highland Council and Highland and Island Enterprise.

White, H. 1973. *Metahistory: The Historical Imagination in Nineteenth-Century Europe*. Baltimore: Johns Hopkins University Press.

Wievorka, A. 1998. *Ere du témoin*. Paris: Plon.

Wievorka, A. 1999. From survivor to witness: Voices from the Shoah, in *War and Remembrance in the Twentieth Century*, edited by J. Winter and E. Sivan. Cambridge: Cambridge University Press, 125-41.

Wievorka, A. and Mouchard, C. 1999. *La Shoah: Témoignages, savoirs, oeuvres*. Presses Universitaires de Vincennes/Cercil.

Williams, A. 1981. *A Heritage for Scotland. Scotland's National Museums and Galleries. The Next 25 Years. A Report of a Committee appointed by the Secretary of State for Scotland under the Chairmanship of Dr Alwyn Williams*. Edinburgh: HMSO.

Williams, R. 1973. *The Country and the City*. London: Chatto and Windus.

Wilson, B. 1997. Still pressing for freedom. *The Herald*, 4 April, 17.

Winter, J. and Sivan E. 1999. *War and Remembrance in the Twentieth Century*. Cambridge: Cambridge University Press.

Witcomb, A. 2003. *Re-Imagining the Museum: Beyond the Mausoleum*. London: Routledge.

Witcomb, A. 2009. Migration, social cohesion and cultural diversity. Can museums move beyond pluralism? *Humanities Research* [Online], 15(2), 49-66. Available at: http://www.epress.anu.edu.au/hrj2009_02_citation.html [accessed 12 December 2009].

Withers, C.W.J. 1988. *Gaelic Scotland. The Transformation of a Culture Region*. London: Routledge.

Withers, C.W.J. 1992. The historical creation of the Scottish Highlands, in *The Manufacture of Scottish History*, edited by I. Donnachie and C. Whatley. Edinburgh: Polygon, 143-156.

Withers, C.W.J. 2005. Landscape, memory, history: *Gloomy Memories* and the nineteenth-century Scottish Highlands. *Scottish Geographical Journal*, 121(1), 29-44.

Wittig, K. 1978. *The Scottish Tradition in Literature*. First published 1958. Edinburgh: Mercat Press.

Womack, P. 1989. *Improvement and Romance: Constructing the Myth of the Highlands*. Basingstoke: Macmillan.

Woodley, P. 2008. Letters to the editor: Timespan – we're proud but not complacent, *Northern Times*, 1 August, 9.

Wright, P. 1985. *On Living in an Old Country*. London: Verso.

Young, J.E. 1993. *The Texture of Memory: Holocaust Memorials and Meaning*. New Haven: Yale University Press.

Youngson, A.J. 1973. *After the Forty-Five*. Edinburgh: Edinburgh University Press.

Index

For Product Safety Concerns and Information please contact our EU
representative GPSR@taylorandfrancis.com
Taylor & Francis Verlag GmbH, Kaufingerstraße 24, 80331 München, Germany

www.ingramcontent.com/pod-product-compliance
Ingram Content Group UK Ltd.
Pitfield, Milton Keynes, MK11 3LW, UK
UKHW020011200926
475702UK00014B/1814

* 9 7 8 1 1 3 8 2 4 6 0 9 6 *